FRICTION, WEAR,
AND **EROSION ATLAS**

FRICTION, WEAR, AND EROSION ATLAS

Kenneth G. Budinski

CRC Press
Taylor & Francis Group
Boca Raton London New York

CRC Press is an imprint of the
Taylor & Francis Group, an **informa** business

CRC Press
Taylor & Francis Group
6000 Broken Sound Parkway NW, Suite 300
Boca Raton, FL 33487-2742

First issued in paperback 2017

Version Date: 20130624

ISBN 13: 978-1-4665-8726-7 (hbk)
ISBN 13: 978-1-138-07431-6 (pbk)

Library of Congress Cataloging-in-Publication Data

Budinski, Kenneth G.
　　Friction, wear, and erosion atlas / Kenneth G. Budinski.
　　　　pages cm
　　Includes index.
　　ISBN 978-1-4665-8726-7 (hardback)
　　1. Mechanical wear--Pictorial works. 2. Materials--Mechanical properties--Pictorial works. 3. Tribology. 4. Lubrication and lubricants. I. Title.

TA418.4.B83 2013
620.1'1292--dc23　　　　　　　　　　　　　　　　　　　　　　　　　　2013021473

This book is dedicated to my beloved wife and partner in all matters

Marilyn

Contents

Preface

This book was first proposed in the ASTM G 2 Committee on Wear and Erosion in 1980 or thereabouts. They suggested a project to catalog photos of different modes of wear and erosion. The committee established a task group and people volunteered to participate. The plan was to catalog the photos and make some sort of publication out of it. Some of us task force members submitted photos, and the effort continued for a year or two and then participation waned. The purpose of the effort was to show readers what different modes of wear look like so that they could identify the cause of failures and take appropriate action to prevent repeat failures. The objective of the book was to lessen the annual cost of wear in the United States and other industrialized countries.

The purpose and objective of this book are still the same. This book shows people with wear problems what mode of wear or erosion predominates in a mechanism or device. There are medical references that show photos of different maladies to teach medical students about different diseases. This book is analogous. No doctor has personally seen people with every disease, and that is why they have compilations of photos of different diseases—to help diagnose. We want this book likewise to be a diagnosis aide.

We started this book as a repeat of the ASTM effort; however, the compilation of wear and erosion photos was made by one person, and the result was a loose-leaf binder of categorized wear and erosion photos—mostly machine failures. It was used as a feature at Bud Labs (Rochester, New York) commercial exhibit at various trade shows.

In 2011, the subject of a wear and erosion atlas reappeared at a committee meeting, and a task group was formed to help convert the Bud Labs commercial exhibit book of photos into a viable reference book. Most of the wear and erosion photos were in the Bud Labs book, and the task group was to serve as reviewers and collaborators to get the reference text published. This book is the culmination of this second ASTM G 2 effort.

The book contains chapters on all of the "popular" wear and erosion modes—those generally agreed to by tribology researchers. We added two chapters on friction. We cannot show photos of different kinds of friction, since friction is a system effect. There are friction problems, as there are wear and erosion problems, but they manifest themselves in forces being too high for a power source to overcome or too low to do a required job—like prevent slipping. What we offer to help readers understand is the various manifestations of friction, such as force traces from a laboratory test rig for a wide variety of test couples. These are intended to give the reader guidance in the use and investigation of friction force outputs of mechanisms.

We start the book with a glossary of the terms that apply to friction, wear, and erosion. The two longest wear chapters, Chapters 3 and 4, address the two most important (from the standpoint of mechanisms) forms of wear: adhesive wear and abrasion. We have chapters on rolling wear, impact wear, and the most important, modes of erosion: liquid droplet, solid particle, slurry, liquid impingement, and cavitation.

Overall, this atlas is the accumulation of examples from 50 years of tribology consulting and research. It is intended to aid and guide engineering students and others who do not have tribology as their specialty. It is offered to help people understand the breadth of the field of tribology. It is offered to help anyone with a friction, wear, or erosion problem. It is not a teaching text, but a reference text—a sort of dictionary of tribology. It shows what abrasion can look like, what adhesive wear can look like, what solid particle erosion can look like, and what rubber versus steel friction force can look like. We hope that it helps readers identify and solve problems, and thus reduce the annual cost of friction, wear, and erosion.

KGB

Acknowledgments

A special thanks to Steven Budinski from Bud Labs, who conducted many special tests to develop data for this atlas, and to Mark Kohler of Arnprior Corp for his graphics help and the cover photo. We also owe a debt of gratitude to the ASTM G 2 Committee on Wear and Erosion for suggesting this atlas and for their efforts in getting this project started.

About the Author

Ken Budinski started his engineering career as a cooperative engineering student at General Motors Institute in Flint, Michigan (now Kettering University). He obtained a BS degree in mechanical engineering in 1961 after five years of cooperative work sessions at Rochester Products Division of General Motors. His fifth-year thesis project was "Thermal Control the Diecasting Process." After GMI, Ken did his graduate work at Michigan Technological University in Houghton, Michigan, and obtained an MS degree in metallurgical engineering in 1963. His thesis was "Effect of Solute Concentration on the Yield Strength of Alpha Iron."

Ken joined Eastman Kodak's Materials Engineering Laboratory in Rochester, New York in 1964. He started as a development engineer and retired in 2002 as a senior technical associate. During his 38-year tenure at Kodak, he specialized in tribology problems. Kodak had many unique problems in this area because of photoactivity concerns. Lubricants could not be used in many pieces of production equipment and in many manufacturing areas. He developed countless material solutions for sanitary and unlubricated sliding systems. His laboratory became the corporate resource for tribology problems and tribotesting. As part of his Kodak work he became significantly involved with ASTM International in developing tribotesting standards and for his tribology research. He authored more than 50 papers in refereed journals and presented more than 100 papers at conferences all over the world.

Ken retired from Eastman Kodak in 2002 and joined Bud Labs, also located in Rochester, New York. Bud Labs is his son's tribotesting company, and he became technical director. He is still in that position today. Bud Labs started as a developer and manufacturer of tribotesting machines, but in 2009 it stopped building testing machines to concentrate on performing tribotests, both standard and nonstandard, for worldwide clients. It performs tests in most areas of tribology, but sees the highest volume of testing in abrasion, sliding wear, solid particle erosion, and friction.

Ken is a fellow in ASTM International, ASM International, and the Rochester Engineering Society, and is currently chair of the ASTM G 02.5 Subcommittee on Friction. He has won many awards for his technical contributions and is the author of a materials textbook, *Engineering Materials: Properties and Selection* (Pearson Education), which is now in its ninth edition. He has authored four other technical books on subjects ranging from technical writing to tribotesting. His *Guide to Friction Wear and Erosion Testing* has been accepted as a significant reference in tribology circles. This atlas is his seminal work in the area of tribology. It includes learnings from his current research into mechanisms of abrasion, galling, solid particle erosion, and friction. Bud Labs performs proprietary research projects, but also maintains continuing research programs in tribotesting and in the mechanisms of material damage with various forms of wear and erosion. This is his current area of research. Tribology has been his life's work, and this atlas is the product of that work.

1

Introduction

The problem/need addressed in this book is the recognition of the various ways that wear erosion and friction manifest themselves in machines, devices, and engineering and science in general. It is about what tribology looks like in the field. As is the case in the health care industry, treating an illness starts with a diagnosis of the malady. This is a critical first step in addressing any health problem. It is also like this in tribology. Solids do not just wear or erode; they do so in many different ways—different mechanisms prevail and different treatments are necessary. The common factor in wear and erosion is progressive loss of material from solid surfaces, but how that occurs is the key to minimizing losses and solving problems that arise for these progressive material losses.

Friction is present in every mechanical device, every moving joint, every place where something slides, flows, or rolls on another substance. Like wear and erosion, there are many types of friction, and dealing with them again requires diagnosis of the type of friction that exists in a system. The common denominator in friction is resistance to motion or flow. How that resistance occurs and is measured depends on the type of friction. Designers need to understand the manifestations of friction and how to deal with it.

What is the importance of friction, wear, and erosion? There are estimates in government documents that place the cost of friction, wear, and erosion at as much as 10% of the gross domestic product (GDP). An incontrovertible example of the annual cost of wear is the automobile. The average life of an automobile in the United States is about 150,000 miles (250,000 km). This mileage may take 10 years for some owners to accumulate, but a salesperson with a wide sales territory may use up his or her 150,000 miles in 2 or 3 years. Whatever the time interval, millions of new cars are sold each year in the United States at an average cost of about $30,000. If yearly sales are 5 million vehicles, the annual cost is $150 billion, assuming that new cars are purchased to replace worn cars. This is the correct assumption in most cases. New drivers come of age each year, but they usually do not start driving with a new car. Commercial vehicles, like trucks involved in interstate shipping, have a life closer to 500,000 miles because they use diesel engines, which have fewer rub strokes per mile than gasoline-powered engines, but their replacement cost is much higher, possibly $100,000. If an interstate truck is on the road for 10 hours a day, 300 days a year, averaging 50 mph, the vehicle will accumulate 150,000 miles in a year, and thus last only about 3 years; the annual cost is about $30,000. And there are ever-increasing trucks on U.S. interstate highways. Then there are the tire and brake costs that go along with the vehicle costs, and all of these costs are largely due to wear. The friction costs that go along with annual vehicle costs are many, but the most obvious is the horsepower losses in engines in overcoming the sliding resistances arising from rubbing parts. These are estimated to be as much as 10% of an engine's total horsepower. Friction, wear, and erosion costs are high, and this is a worldwide phenomenon. In some cases, tribology problems are limiting factors. That is, friction, wear, or erosion limits what can currently be done with a process, device, machine, etc. For example, drill head wear is the limiting factor in drilling oil, gas, or water wells. A head may only last to 100 meters and need replacement. The life of most automobile brakes is less than 50,000 miles. This is the current technology

TABLE 1.1

Applications Where Friction, Wear, or Erosion Limit the Economics, Successful Functioning, or Service Life

Wear Is a Limiting Factor In:	Friction Is a Limiting Factor In:	Erosion Is a Limiting Factor In:
Vehicle life	Vehicle fuel use	Jet engines
Tire life	Tire traction	Helicopter rotors
Shoe life	Bearings	Dams
Clothing life	Ship fuel use	Waterways
Machinery life	Railroad wheel traction	Hydroelectric facilities
Tool life	Human joints	Human joints
Mineral extraction	Furniture coverings	Steam handling
Mineral processing	Clothing (feel)	Steam turbines
Tillage tools	Hair care (feel)	Slurry handling
Well drilling tools	Walkways/footwear	Particle handling
Control valves	Medical insertion devices	Exhaust fans
Etc.	Etc.	Etc.

barrier, but 40 years ago that limiting number was less than 30,000 miles. So these wear limits can be improved, and hopefully some of the information in this atlas will assist in that happening. Table 1.1 is a tabulation of some instances of friction, wear, and erosion being limiting factors.

How can this atlas help reduce the annual cost of wear to manufacturers, consumers, and governments? The objective of this book is just that: Reduction of the annual cost of friction, wear, and erosoin. The purpose of the book is to show readers what friction, wear, and erosion look like and what causes them, so that they can mitigate their effects in design and material selection. Categorizing the type of tribological problem that can exist in a machine or mechanism is step 1. It is the diagnosis step. As in medicine, diagnosis is key. Is a tool, part, gear, cam, etc., going to deteriorate in service by abrasion, by rubbing wear, by tribocorrosion? This must be established before a solution is sought. A doctor cannot cure a cough unless he or she identifies the cause. Is it a cold? Is it an allergy? Is it emphysema? Is it pneumonia? A part is wearing. Is it abrasion? Is it fretting corrosion? Is it galling? This book has a chapter on each major type of wear and erosion. Photos are presented to show the appearance of affected surfaces, macroscopically and microscopically. How is the material removed from a surface? There are two chapters on friction: one dealing with sliding friction, and the other dealing with rolling friction. Friction is a force, so we cannot take a photo of it, but we present force readouts from instruments for the various kinds of fiction that can be encountered. The book ends with three chapters on how to address tribology problems. There is a chapter on materials that can be used to solve wear problems in their as-purchased or heat-treated condition. We use the term *bulk materials* to describe them. Then, there is a chapter on surface engineering that discusses the coatings and treatments that can be applied to bulk materials to improve their tribological behavior. The final chapter is a guide on how to go about selecting materials for tribological applications. There are eight appendices that contain information that also may be helpful in dealing with tribology problems. The largest appendix deals with fusion hardfacing, which is an extremely useful tool in dealing with many forms of wear and erosion.

Like it or not, every designer must ask himself or herself on every design: How long do I want this xxx to last? What is its design life? Is it 1 year, one use, 1 million uses, 10 years? Followed by: How might this xxx fail? Will it short out? Will it corrode? Will it wear? Will erosion be a factor? Will high or low friction keep it from doing its job? If the object under design has moving parts or will be rubbed on, or is subject to liquid or particle impingement, chances are that tribocharacteristics of materials will be a design concern, and this atlas can help designers deal with these anticipated problems. Possible tribological problems are ignored in far too many designs. We have all witnessed massive vehicle recalls for friction or wear problems. Entire buildings have had to change out new flooring because some designer ignored the slip resistance of that flooring under tracked-in slush conditions. Unanticipated cavitation damage to massive concrete structures has shut down hydropower plants. Tribology cannot be ignored in design, and this atlas can raise the awareness of designers to potential tribology concerns in a design.

2

Glossary of Tribology Terms

There are entire books and encyclopedic directories published by various organizations on wear/erosion terminology. ASTM has standards for these terms (G 40 and D 4125), but there are no overarching review processes to ensure that the definitions from different organizations or societies agree. We will present our definitions. They may or may not concur with other published definitions that are based upon all of the definitions that we are aware of, our own examples of worn/eroded surfaces, and observations in performing laboratory tribotests.

Figures 2.1 and 2.2 show the most prevalent forms or modes of wear. However, before we go any further, let us define *mode* and *form*.

$$\left\{ \begin{array}{l} \text{Form (n), a type/variation of something, a shape, a business or legal document} \\ \text{Mode (n), a type/variation of something} \end{array} \right\}$$

Academic literature often uses the word *mode* to identify a prevailing triboprocess, but *form* also does the job. *Mode* sounds more scientific. In most cases, this book will opt for *mode*.

Next, we need to define wear and erosion. Both wear and erosion require material removal; both require that the material loss is progressive. As mentioned previously, erosion requires a contribution to the damage from a fluid, while wear does not. Both wear and erosion require fracture of material from a surface. There are only two common ways to remove material from a surface: (1) fracture it and (2) dissolve it. All machining operations are a form of fracture. Corrosion processes remove material by dissolving it. So with these definition requirements in mind, we offer these definitions for wear and erosion.

General Tribology Terms

Wear (n): Progressive loss or damage of a solid surface caused by forcible sliding contact with another solid.

Erosion (n): Progressive loss or damage to a solid surface caused by contact with a moving fluid. The fluid can be single phase or multiphased, and the material removal can include a corrosion component.

Tribology (n): The art and science of reactions and devices associated with rubbing contact between solids or fluid motion in contact with solids (e.g., friction, wear, erosion, lubrication, wear/erosion mitigating devices, etc.).

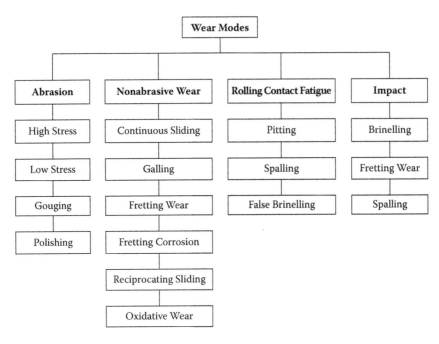

FIGURE 2.1
Wear modes.

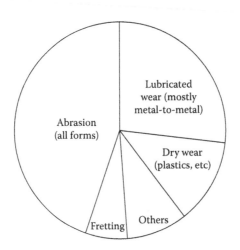

FIGURE 2.2
Estimated importance (based upon research activity/spending) of various modes of wear.

The *tribo* prefix has been added to many other words by the people working in the field (tribologists):

Tribotesting (n, v): Laboratory studies and devices associated with wear, friction, erosion, and lubrication.

Tribocorrosion (n, v): Wear and erosion processes wherein chemical attack is known to play a substantial role in causing material loss or damage.

Tribometer (n): A testing device used for wear, friction, erosion, or lubrication studies.

Tribological (adj): Relating to tribology.

There are probably a dozen additional "triboterms" in the literature, and they all mean something associated with friction, wear, and lubrication. The word *tribology* was coined in the 1980s in the UK to make a single word for friction, wear, lubrication, and erosion activities and damage. The intent was to make a more powerful and inclusive term to increase public awareness of the field and government funding for research in these areas.

Unfortunately, after all of these years of being in UK dictionaries and encyclopedias, it is not in most online dictionaries, and only 0.001% (or some other low number) of the general public know the term. So words with a *tribo* prefix are used only by those working in the field. However, we encourage readers of this book to use triboterms. It is reasonable to talk about wear/erosion processes as triboprocesses because wear, friction, lubrication, and erosion may all be significant contributors in problems associated with rubbing. The *tribo* prefix is inclusive and thus desirable.

Getting back to necessary definitions, it may be well to show the relative importance of the various modes of wear, friction, and erosion. Figure 2.2 is an estimate of the relative research emphasis connected with, and the economic significance of, various wear modes. Figure 2.4 presents the same importance ranking for types of erosion. These are opinions because there is no entity (in the United States) that compiles cost data on triboprocesses, but in the testing business, abrasion testing is requested far more than any other form of wear or erosion.

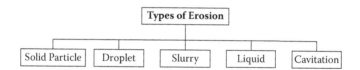

FIGURE 2.3
Types of erosion.

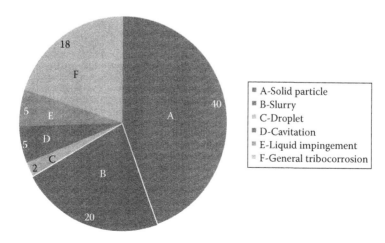

FIGURE 2.4
Estimated importance of various types of erosion.

Abrasive Wear Terms

From the cost standpoint, it is well documented that every ton of coal mined produces a known amount of tool costs, and this situation exists for all minerals extracted from the earth. The cost of abrasion in mining could very well be hundreds of billions of dollars annually. Then there is the tire abrasion. The worldwide tire industry may be $50 billion a year industry, and a good portion of that cost is replacement for abraded tires on older automobiles and trucks. Another obvious high-cost wear mode is lubricated wear in vehicles: engine components, bearings, pivots, etc. The life of an engine is determined by the amount of wear that has accumulated. Replacing a worn-out engine usually requires replacement of the entire vehicle at significant cost. Dry wear, as in brakes, costs the average automobile owner about $1000 in the life of a vehicle, and there were 200 million vehicles licensed in 2012 in the United States. These examples are presented to support our relative importance estimates.

Shoreline erosion by bodies of water (rivers, lakes, streams, oceans) is unquestionably the most costly form of erosion. Every hurricane or cyclone that comes ashore on populated land can easily create tens of billions of dollars damage, thus making liquid erosion the most costly mode of erosion. This form of erosion can be controlled to some extent, but it is not, mostly because governments often do not allow erosion control devices on shorelines for reasons usually not based upon science. However, liquid erosion occurs in every pipeline conveying a liquid, and thus the cost of replacing pipelines and control devices that have become a concern because of liquid erosion damage is huge globally. For example, every leaky toilet probably became leaky because of liquid erosion of a shutoff seal.

We would be remiss if we did not mention at some point that the "corrosion community" has laid claim to most modes of erosion since most involve a corrosion component. In fact, corrosion engineers have contributed greatly to solving erosion problems by using electronic corrosion measuring techniques such as potentiodynamic and potentiostatic polarization to quantify the corrosion contribution in various erosion processes, such as liquid impingement and slurry erosion. There is not a problem with two groups of specialists, the tribologists and corrosion engineers, claiming research rights to erosion. There is more work to be done than there are researchers in both communities. Now, let us define the wear and erosion modes listed in Figures 2.1 and 2.3.

Abrasion (n): A form of wear characterized by progressive removal of material from a solid surface by the action of forcible sliding contact with hard substances or protuberances.

Low-stress abrasion (n): Progressive removal of material from a solid surface by the forcible sliding contact with hard substances (abrasives) or protuberances with contact stress levels generally below the compressive strength of the abrasive and the counterface.

Counterface (n): Typically the stationery surface in a wear solid-on-solid test couple. For example, if a ball is sliding on a flat surface, the flat surface is called the counterface and the ball is called the rider.

The words *counterface* and *rider* are tribology jargon, but they have utility in writing about wear testing. Unfortunately, it is not common to use the term *counterface* for erosion.

High-stress abrasion (n): Progressive removal of material from a solid surface by the sliding contact of hard substances with forces sufficient to cause fracture/crushing of the hard substance.

Gouging abrasion (n): Progressive removal of material from a solid surface by sliding contact and impact of hard substances with sufficient force to cause plastic deformation in the shape of craters and the like in the surface subjected to the sliding and impacting substances. Crushing rocks usually produce this type of abrasion on the crushing and handling equipment.

Polishing abrasion (n): Progressive removal of material and lowering of surface roughness on a solid surface by the sliding of fine hard substances on compliant materials under forcible contact with the surface. This form of wear could be assisted by a conjoint chemical reaction such as corrosion.

Polishing particle (n): Particles one micrometer in diameter and finer embedded in cloths or other materials that allow embedment of the particle. The particles are usually hard substances like silica, alumina, silicon carbide, etc.

Polishing of surfaces for electronic devices like integrated circuits requires precision polishing, and they tend to use oxide particles embedded in polymers such as polyurethane foams and liquid facilitators that may chemically react with the surface being polished. Often, the liquid is designed to oxidize the surface and the lap polishes mostly by continuous removal of the oxidized material. Chip manufacturers call their process planarization or chemical-mechanical planarization (CMP).

Handrails, door knobs, and similar things in public places often become "polished" by the abrasive action of substances on people's hands, with possible contribution from chemical attack caused by uric acid and other "chemicals" on people's hands. It is chemical-mechanical polishing.

Electropolishing (n): The removal of a material from a solid surface by electrochemical action in such a manner that it lowers surface roughness.

Electropolishing is not as widely used as much in mechanical or chemical-mechanical processing because every material requires a suitable electrolyte, voltage, current density, cathode material, etc. These process requirements are often not readily available. For example, there are electrolytes known to produce polishing on stainless steel, but those electrolytes and conditions may not work at all on carbon steel or other materials. The use of chemical assist in polishing often requires a development project, while polishing with 1 μm aluminum oxide on a soft cloth polishes most surfaces with continued rubbing.

Polishing (n): The process of lowering the height parameters of the surface texture of a solid; a synonym for polishing abrasion.

Surface texture (n): The microscopic geometric features that exist on solid surfaces (surface texture includes roughness, waviness, and other mathematical descriptors of surfaces, like the one shown in Figure 2.5).

Surface finish (n): The surface texture intentionally produced on a solid surface in manufacture or treatment of the surface.

Roughness (n): The average peak-to-valley height of the microscopic features on a solid surface (Figure 2.6).

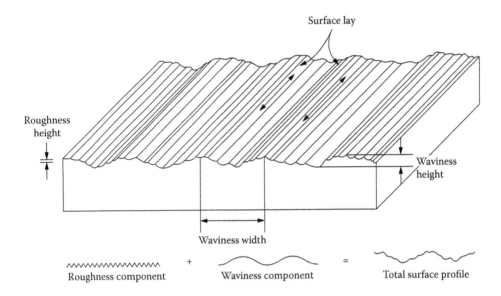

FIGURE 2.5
Components of surface texture.

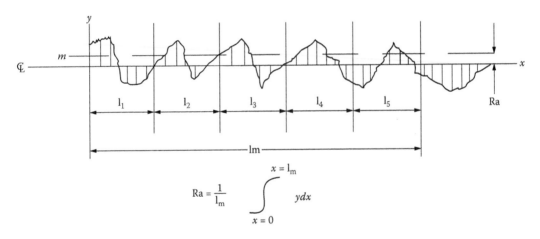

$$Ra = \frac{1}{l_m} \int_{x=0}^{x=l_m} y\,dx$$

FIGURE 2.6
Surface texture as characterized by the surface texture parameter Ra, roughness average. It is the arithmetic average of all distances of the surface profile from the centerline within a specified measurement length (lm).

> **Waviness** (n): The surface height variations that occur in periodic form and have spacings that are large compared to the spacing of surface roughness peaks and valleys (waviness widths are typically in millimeters and wave heights in micrometers).
>
> **Asperities** (n): Tribology jargon for surface roughness peaks.

There are many wear and erosion models based upon the asperities of one surface interacting with the asperities on a contacting solid surface. An asperity in such models is usually assumed to have a cone shape with a certain base diameter and cone (peak) height from some mean reference line that defines the surface. When one solid surface is placed upon another solid surface, the real area of contact is first determined by errors of form and waviness of both contacting solids. For example, two metals that obtained their surface texture by surface grinding will contact each other at mating waveforms. However, those waveforms also contain additional surface features, like fine scratches and "burrs" left by the fracture of the surfaces by the grinding abrasives. These finer features are the first to carry contact load, and thus they would plastically deform (flatten) at the contact points. These fine surface features could be construed as the asperities in wear models, but in no way do these asperities carry significant load over the apparent area of contact intact. Real area of contact is determined primarily by surface waviness and errors of form. If a flat plate found to have a crown of 0.1 mm is placed on a flat surface with out-of-flatness less than 0.01 mm, the convex shape will have a line contact and the real contact areas within that line are determined first by the waviness of the line contact and then by the roughness of the line contact. Real surfaces and real contact areas are not made up of atomically flat planes sprinkled with perfect cones (asperities), and thus we are of the opinion that the real area of contact between conforming surfaces is determined by waviness and error of form interactions rather than asperity height and base contacts.

Surface texture parameters (n): Mathematical descriptions of the departures of a surface from perfect flatness and planarity. There are at least 20 in common use: root mean square (rms), mean peak spacing, 10-point height, etc.

There are countless technical papers that correlate wear and friction results with surface texture parameters. This may happen only because there are so many available parameters. If one does not correlate, a researcher can continue on down the list of 20 until one does.

Abradant (n); A substance that produces abrasion on a solid surface.

Abrasive (n): A substance with known ability to produce abrasive wear.

Nonabrasive Wear Terms

Nonabrasive wear (n): Progressive loss or damage of material from rubbing solid surfaces produced by interactions between the contacting surfaces and not by abrasive substances intentionally introduced into the tribosystem.

Nonabrasive wear is a euphemism for adhesive wear. There are a number of wear modes that probably initiate by adhesive interactions between the rubbing surfaces, and thus they should be listed under a header called adhesive wear. However, it is well known that metal-to-metal tribosystems often produce wear debris that oxidize by repeated rubbing, and these metal oxides can be abrasive to the rubbing surfaces and may contribute to the material removal. Thus, adhesive wear can become abrasive wear. Thus, it became common practice to call these galling, solid-on-solid, and fretting wear nonabrasive wear modes rather than adhesive wear modes.

Adhesive wear (n): Progressive loss of material from or damage to a solid surface in rubbing contact caused by solid-state bonding (adhesion) between the rubbing surfaces.

The rubbing-induced bonding/adhesion usually starts with local areas of transfer (transfer film). The transfer can become a continuous film, wear debris, platelets, or microscopic mounds adhering to the contacting surface. The damage to the prevailing surface textures of rubbing surfaces can be as slight as a microscopic change in surface texture, or it can be significant change to the surface texture, such as galling. The damage can be confined to one of the contacting surfaces or it can be on all contacting surfaces.

Scoring (n): Wear-produced furrows/scratches in a prevailing surface texture on a solid surface subjected to rubbing contact by a conforming solid. The damage aligns with sliding direction. The furrows/scratches can occur on one surface or more of the surfaces that rub.

Scuffing (n): A synonym for scoring-localized wear/change in surface texture aligned with the direction of sliding.

Seizure (n): The unanticipated stopping of motion between sliding members caused by rubbing-induced adhesion between the rubbing surfaces.

Fretting (n): Small amplitude (10 to 300 µm) relative sliding between surfaces in forcible contact.

Fretting corrosion (n): Progressive loss of material or damage to a solid surface caused by small-amplitude relative sliding when in forcible contact with another surface under environmental conditions that cause the rubbing surfaces to oxidize or chemically react; thus environmental reaction contributes to material loss/damage.

Fretting wear (n): Progressive loss of material from a solid surface caused by small-amplitude relative motion with a contacting surface.

The reason there are lower limits on what slip amplitude can cause fretting wear and fretting corrosion is that it is thought that below some critical slip amplitude, the contacting surface features elastically deform to accommodate the relative motion. There is no rubbing, and thus no wear. Conversely, if the slip amplitude gets too large, the "special" things that occur on surfaces no longer occur and normal reciprocating wear occurs: *gross-slip* is the term applied. The upper limit is also in contention. Some researchers believe it to be 300 µm, and others as low as 100 µm. The ASTM test standard G 204 uses a slip amplitude of 50 µm, and that means total measured displacement. This would be the product of one complete revolution on a device driven by a rotary motor.

Fretting amplitude (n): The total relative motion between two contacting surfaces in a cycle.

It is common to use a ball-on-flat configuration in performing laboratory fretting tests, and this contact geometry leads to an annular scar on the flat counterface when the slip amplitudes are low (line < 10 µm). The center of the donut is called the no-slip zone, and, of course, there is a partial slip zone and a gross slip zone where wear "scratches" are evident. Most contacting metal couples that experience fretting motion damage by fretting corrosion. There is reaction of the rubbing surfaces with the environment if the environment can oxidize. Fretting damage of metals in air is often characterized by metal particles that oxidize with repeated rubbing, and thus there is a corrosion component to the damage. It is usually substantial. Fretting wear rather than fretting corrosion is common on plastic/plastic couples because plastics do not oxidize like metals in repeated rubbing. The

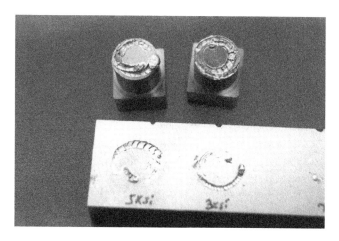

FIGURE 2.7 (See color insert.)
Excrescence formation on test specimens in the ASTM G 98 galling test. This damage occurred in one 360° rotation of the upper button on the lower counterface or block. The left test was performed at an apparent contact pressure of 5 ksi (35 MPa) and the right at 3 ksi (21 MPa).

damage produces plastic particles rather than an oxide that can be abrasive. Thus, the term *nonabrasive* is used for fretting wear and corrosion even though there may be abrasion from hard substances like oxides.

Galling (n): Severe form of adhesive wear characterized by localized formation of excrescences (material flowed up from one or both surfaces) that can inhibit sliding of a conforming surface and lead to seizure (Figure 2.7).

Gall (v): The act of galling.

Adhesive wear (n): Progressive removal of material from a solid surface when forcibly rubbed by another solid surface. It is often characterized by localized material from one surface transferring to the other by solid-state bonding. The transferred material may transfer back or form a transferred "coating," or it may form detritus.

Adhesive transfer (n): Material from a solid surface that has been solid-state bonded to another solid surface in forcible rubbing contact. The transferred material can be atomically thin, microscopic, or macroscopic. It can be semicontinuous or in clusters, plastic particles, or other such forms.

Rolling contact fatigue wear (n): Progressive loss of original material from one or more solid surfaces caused by repeated compressive stressing of the contacting solid surfaces by a rolling element.

Probably all forms of wear involve fatigue, cyclic application of stress. So, all wear and erosion processes could fit into a category of fatigue wear; therefore, we added *compressive* to the state of stressing. The most important types of fatigue wear occur in tribosystems involving Hertzian contacts, such as a sphere on a flat, a flat on a cylinder, a cylinder on a cylinder, crossed cylinders, etc.—all contacts that start as point or line. This means all rolling element bearings, gears, rolling cams, and the like. In these systems, the highest stress on the contacting surfaces occurs below the surface and wear particles often originate as subsurface cracks produced by repeated rolling or impact.

Impact wear (n): A form of fatigue wear characterized by progressive loss of material from a solid surface by cyclic impact normal to a surface with no intentional tangential sliding.

Pounding steel stakes in the ground with a steel hammer will subject the hammer head to impact wear. There is relative sliding caused by elastic or plastic strains at points of contact.

Oxidative wear (n): A form of mild metal-to-metal wear characterized by removal of metal from one or both surfaces in particulate form, and the particles and surfaces react with oxygen in the environment to form metal oxide wear debris. The metal surfaces become covered and separated by the debris. In ferrous systems, the oxide usually looks like rust.

Oxidative wear is not a widely used term, but is useful to describe what happens when metals rub on each other under moderate normal forces in the absence of an intentional lubricant. Most door hinges are assembled without a lubricant on the hinge pin. Most liquid lubes will leach out and damage painted surfaces. So carpenters do not lubricate the pins, and after several years of use, the pins are covered with what appears to be rust. It is wear debris that started out metallic, but with repeated friction and rubbing, the freshly cleaved particle surfaces oxidize and become iron oxide when the hinge pins are steel. The oxide that is trapped between the rubbing surfaces separates the metals that started out touching and the particles act as a lubricant to separate the surfaces and keep the wear rate low.

Fatigue (n): Cyclic application of stress to a solid material.

Fatigue can lead to cracking (surface or subsurface), and cracks can progress to cause fracture of particles or fracture of pieces from large components. Fatigue damage can come in a few cycles or large numbers (10^6, 10^{10} cycles, etc.)

Spalling (n): The ejection of material in the form of particles, platelets, pieces of coatings, and the like from a solid surface.

In tribology, spalling often occurs when rolling members in point or line contact cause subsurface fatigue cracks. When the subsurface crack propagates all around to the surface, a spall occurs. The piece ejected is a spall. When lots of these occur, spalling exists. Hard, thin coatings can spall from the compressive stresses of balls or rollers rolling on the coating. Cracks occur in the bond zone of the coating to the substrate. Spalling frequently occurs in concrete surfaces subjected to deicing salts in freeze-thaw zones. In this instance, the spalling is due to stresses caused by repeated formation of ice within the concrete and repeated thawing. The stresses produce surface cracks and pieces are ejected. Most are macroscopic.

Brinelling (n): Hemispherical or cylindrical depressions on a solid surface caused by static loading of balls or rollers on another surface with sufficient force to cause plastic deformation of that surface.

Brinelling is a form of wear even though no material is removed when, for example, a ball is pushed into a surface with sufficient force to cause a "dent" and the mating surface is damaged; thus, wear is said to occur. However, false brinelling is a form of wear that does involve material removal from a solid surface. False brinelling is surface damage at Hertzian contacts caused by fretting motion. The classic example often cited for false brinelling is automobiles in rail or truck transit. They are not moving, but the ball and roller bearings in the auto are moving slightly due to the transport vibrations. If the vibration is sufficient, the balls and rollers in the auto's bearing can slip relative to their conforming surface and fretting wear or corrosion can occur. This was a problem in the past, which has been solved by reduced motion and lubricants that mitigate fretting damage. The observable "dents" from false brinelling will be seen as typical fretting damage cavities/pits.

Erosion Terms

Solid particle erosion (n): Progressive removal of material from a solid surface by repeated strikes of solid particulate matter that is harder than the target surface.

If the particulate is soft with respect to the target surface, no damage will occur. For example, sawdust produced by rotary saws usually produces no damage (other than paint removal) on surfaces that are impacted by the particles. Plastic particles are often impacted in cutting clean particle boards on a table saw, but the particles do not damage the metal target surface. If particle boards are used, which might contain metal chips and many other types of nonwood particles, the situation is much different.

Particle size also plays a role in solid particle erosion. If the size (mass) is very low (like submicron), the particles are likely to not damage a target surface. Some particles tend to stick to a target surface. When this happens, the surface is protected because the impinging particles hit themselves.

Velocity affects particle impingement erosion to at least a squared magnitude. Hard particles like aluminum oxide can be "dropped" on a target surface from a moderate height (10 cm or so) with no damage to the target. There is insufficient velocity to cause damage. Thus, solid particle erosion is a function of particle hardness, mass of particles of particles impacting a target, and particle velocity to the second to fifth power.

Liquid impingement erosion (n): Progressive loss of material from a solid surface caused by the force of a liquid directed at that surface.

Liquid impingement may be conjoint with corrosion. For example, materials like stainless steel are protected from corrosion by passive oxide films. If the impinging liquid is energetic enough, these films can be mechanically removed, allowing attack of the surface that lost its protective surface. Liquid in the form of steam can cause this type of erosion.

Droplet erosion (n): Progressive loss of material from a solid surface caused by the mechanical action of impinging drops of liquid, usually at high velocity.

Droplet erosion will take place on solids that are impacted with rain coming off a roof. Concrete erodes by progressive removal of the concrete binder, leaving the aggregate standing proud. However, high-velocity droplet erosion is of great concern in aircraft. A jet at 600 mph going through a rain field can encounter severe erosion tendencies, particularly on nonmetal components like plastic windshields, radar domes, and even paints.

Slurry erosion (n): Progressive loss of material from a solid surface caused by the mechanical action or tribocorrosion produced in sliding contact with a liquid containing entrained particles.

The particles can be in any size or concentration, and this form of erosion can also have a component of material removal caused by corrosion. Mining operations typically involve handling of slurries, and this type of erosion is commonly encountered.

Cavitation (n): The dynamic generation of bubbles or voids in a liquid, submerged local bubbles that contain no liquid—usually they contain a gas.

Cavitation erosion (n): The progressive loss of material from a solid surface resulting from energetic jets that are produced by collapsing bubbles.

The "jets" that can impinge on a surface when bubbles collapse can produce extreme pressures (like 20,000 psi or 138 MPa) that have sufficient force to remove material. Again, there can be a corrosion component to the material removal.

CMP (n): An acronym for chemical-mechanical polishing.

Chemical-mechanical polishing (n): A form of abrasion involving intentional removal of chemically induced reaction product films on solid surfaces.

This process is widely used to polish semiconductor materials and deposited layers on substrates used to make computer chips. It is more of a manufacturing process than a form of wear or erosion. It is also called planarizing, or chemical-mechanical planarizing.

In summary, step 1 in dealing with tribology issues is to become conversant with the terms used in the business. This glossary contains only a limited number of the terms used in tribology literature, but the end-of-chapter references can be consulted to dig deeper.

Related Reading

ASTM D 4125, *Standard Terminology Relating to Petroleum, Petroleum Products, and Lubricants*, West Conshohocken, PA: ASTM International.

ASTM G 40, *Terminology Relating to Wear and Erosion*, West Conshohocken, PA: ASTM International.

Bayer, R.G., *Mechanical Wear Fundamentals and Testing: Revised and Expanded*, Boca Raton, FL: CRC Press, 2004.

Hutchings, I.M., *Friction and Wear of Engineering Materials*, Boca Raton, FL: CRC Press, 1992.

Kragelski, I.V., *Friction and Wear*, London: Butterworth, 1965.

Stachowiak, G., *New Materials Mechanism and Practices*, New York: Wiley, 2005.

Vajdas, C., Harvey, S.S.K., Lusz, E. *Encyclopedia of Tribology*, Oxford: Elsevier, 1990.

Williams, J.A., *Engineering Tribology*, Oxford: Oxford University Press, 1994.

3

Adhesive Wear

Adhesive wear was defined in Chapter 2 as "progressive loss of material from or damage to a solid surface in rubbing contact caused by solid-state bonding (adhesion) between the rubbing surfaces." Examples of adhesive wear that everyone is familiar with, but may not recognize, are chalk on a blackboard and writing with a pencil on paper. In the former, chalk is transferred to the smooth and harder-than-chalk blackboard. The blackboard can be made from slate or any rigid material with a nonshiny, unlubricated surface. In the latter, soft graphite (lead) is transferred to smooth but porous paper. If a pencil mark is observed under a microscope (Figure 3.1), it will appear black, as platelets deposited on the peaks of the interlocked fibers that constitute a paper surface.

Chalk on slate looks similar—platelets of the softer of the contacting bodies transfer to the harder. These examples are clearly visible to the unaided eye. Often, the transfer film is not. Sometimes it can be atomic—a few atom layers. Sometimes transfer is just hard to detect. This situation often exists with plastics. For example, rubbing most solid surfaces with Teflon® (polytetrafluoroethylene (PTFE)) will produce a significant transfer film that is invisible to the unaided eye, but not to analytical tools like Fourier transform infrared (FTIR) chemical analysis. A classic example of adhesive transfer of one metal to another is to scrape the edge of a U.S. copper penny on a smooth stainless steel surface, then strike a welding arc on this area where the penny was rubbed. The weld will crack. Copper causes hot shortness in molten steels. Hot shortness means cracking on cooling from the molten temperature. This example has been used for many years in training welders not to let their copper connectors scrub or rub on areas to be welded. The transfer may be only atoms thick, but it is there and it can affect the weld.

Another classic example of adhesive transfer is to identify titanium by rubbing the corner of a shape on window glass. If the metal in question is titanium, it will leave a silver streak on the glass. This is an old shop trick to tell stainless steel from titanium when stock may be unmarked or mismarked. Titanium is a very reactive metal, and it has a strong tendency to adhere to other surfaces in rubbing contact (Figure 3.2). Stainless steel and aluminum cookware often leave "scrub marks" when dropped on glass-enameled sinks. This example of adhesive wear is visible in many kitchen sinks in the United States where glass-coated sinks are very popular. In lubricated systems, mating surfaces are separated by a film of lubricant. However, if the operating conditions are not suitable, the surfaces in sliding contact can lose thin-film separations and sliding contact can occur. Adhesion can also occur during these contact periods (like at powering up and powering down a machine), and the result is often one or both materials becoming a constituent in the tribofilm that separates the surfaces. Analytical techniques such as energy-dispersive x-ray analysis (EDAX) and electron microprobe analysis will confirm adhesive transfer in the tribofilm on one or both surfaces.

In summary, adhesion between rubbing surfaces is very common. Soft materials are the most prone to adhesion, and adhesion between surfaces is often the start of the wear process. The ASTM G 2 Wear and Erosion Committee calls wear that starts with adhesive wear nonabrasive wear since once adhesive wear starts, the wear debris or transferred

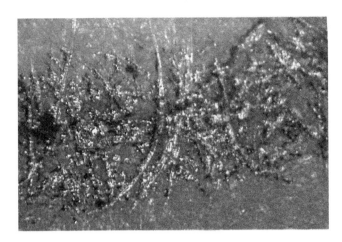

FIGURE 3.1
Adhesive transfer of carbon graphite to paper in writing on it with a pencil (100×).

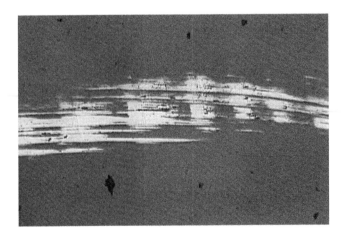

FIGURE 3.2 (See color insert.)
Adhesive transfer of titanium to glass by one forcible rub (100×).

material can react with the environment to form oxides or other wear products that can be harder than the rubbing members, and thus produce abrasion. Adhesive wear can be conjoint with abrasion.

The Mechanism of Adhesion

All surfaces are composed of atoms in either crystal form, as in most metals and ceramics; amorphous form, as in some coatings and treatments; or molecular form, as in plastics. When surfaces make contact with each other, the films that naturally occur on surfaces often separate the atoms or molecules so there is no tendency for adhesion. However, if the force pushing contacting surfaces becomes sufficient locally at spots in the real area of contact, films can break down and the atoms or molecules of the mating surfaces can make

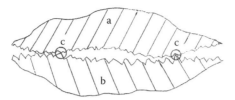

FIGURE 3.3
Adhesion loci are produced by sites of real contact between conforming surfaces (a) and (b).

atomic contact (Figure 3.3) and atomic bonding can occur, depending on the nature of the atoms involved. Of course, like atoms (e.g., steel–iron on steel–iron) will want to bond to each other and adhesion is more likely. Atomic or molecular bonding is less likely with dissimilar atoms or molecules (e.g., steel sliding on a phenolic thermosetting (PF) plastic like countertop laminate). Thus, the mechanism of adhesion between rubbing surfaces is atoms of one surface in atomic contact with atoms of the contacting surface at the waveforms that make up the real area of contact.

Role of Speed, Load, Sliding Distance, Temperature, etc.

Separating films play a significant role in adhesion between surfaces, but so do the operating conditions. Speed of contacting, the loading between members, how much sliding occurs, and environmental factors, such as ambient atmosphere and temperature, affect the tendency for rubbing surfaces to adhere. The Archard equation:

$$W = \frac{kLD}{H}$$

where:
  W = wear
  L = load (force) pushing the sliding members together
  D = sliding distance
  H = hardness of the softer material in the contacting couple
  k = a proportionality constant that is experimentally determined

This equation has been modified and analyzed in every imaginable way over the decades since publication, but most of it survives because it is intuitively correct. Since increased force between members will make adhesion more likely, the more rubbing that occurs (distance), the more likely adhesion will occur, and the harder the material's rubbing, the less the likelihood that one will rub off on the other.

Researchers in this millennium often add elastic modulus in some form to the denominator of the basic Archard equation:

$$W \approx \frac{kLD}{Hf(E)}$$

where E is the elastic modulus of the softer material.

Support for this addition can be found in the use of cemented carbides in tribosystems. They are more wear resistant in sliding systems than probably all metals and ceramics, and they have the highest elastic modulus of commonly available engineering materials. The k in the Archard equation is often used as a metric for the wear of a tribosystem. For example, a hard metal on a hard metal in oil might have a k of 10^{-7}, while a cast iron on steel in oil may have a k of 10^{-6}. The k is called wear factor, wear coefficient, or other names, and it has units derived from the Archard equation.

$$k = \frac{\text{Volume (V)}}{\text{Force (F)} \times \text{Distance (D)}}$$

$$k = \frac{m^3}{Nm}$$

Where: Volume (V) is from a wear test, conducted with Force (F) for sliding distance (D).

It is unfortunate that k is experimentally determined since this means that the adhesive wear model is still not widely used to predict the success of a particular mating couple. Testing is still common in determining the efficacy of a particular mating couple.

Appearance of Adhesive Wear

We showed what adhesive wear looks like in a variety of soft-on-hard couples. What does it look like with common engineering materials?

Galling

We defined galling as a severe form of adhesive wear between mating solids characterized by microscopic material flowing up from one or both surfaces, often coupled with adhesive transfer between surfaces. The operative term in galling is *flowing up* from a surface. Galling looks like Figure 3.4. The up features often lead to seizure, as in the example in Figure 3.4. The shaft was in a hole with not much clearance. The material flowing up from the surface, the excrescence, used up the clearance and galling resulted on shaft removal.

Incipient Galling

Sometimes a material couple produces only microscopic excrescencies, as shown in Figure 3.5. The surface damage is the same as in galling, but more than the naked eye is necessary to see the damage. Usually a 7X loupe is sufficient. Hard metals often exhibit incipient galling. Gear teeth with moderate hardness are prone. There is a standard test for galling: ASTM G 98. This test uses a 360° rotation of a ½ in. diameter flat-ended pin (button) on a flat counterface (block). A test is conducted at a trial load to see if galling occurs. If it does, the load is lowered and the test is repeated until galling no longer occurs. The apparent stress that the couple can tolerate without galling is called the threshold galling stress. Incipient galling and adhesive transfer (Figure 3.6) are not called galling, but they still denote an unacceptable mating couple.

FIGURE 3.4
Galling on a steel shaft that was pressed out of a steel block.

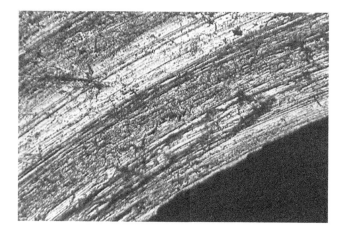

FIGURE 3.5 (See color insert.)
Incipient galling (100×)—microscopic excrescences instead of macroscopic ones.

Scoring

Many times in a galling test one or both surfaces will exhibit plastically deformed furrows. These marks are termed scoring (Figure 3.7), and they usually are from material from one surface that is either an excrescence or transferred material, and it acts as a tool on sliding to create a groove or furrow in one or both mating members. Adhesion between surfaces is the cause.

Scuffing

When scoring becomes profuse, it is often called scuffing. It is common to use this term for the onset of wear in lubricated systems. If the lubricant is doing its job, the mating surfaces retain their original surface texture. Only minor lowering of surface roughness or peak height occurs with time. However, if something unanticipated occurs, such as depletion of additives in the oil, scuffing can occur (Figure 3.8); profuse scoring caused by adhesion

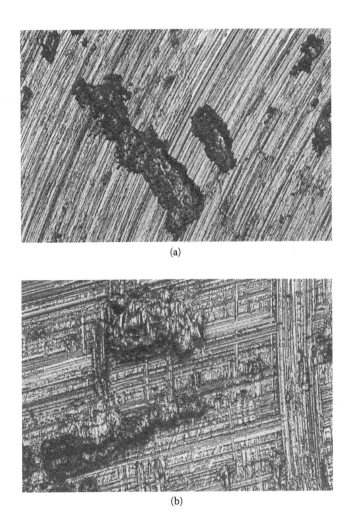

(a)

(b)

FIGURE 3.6 (See color insert.)
Adhesive transfer of titanium (a) to steel (b) in an ASTM G 98 galling test (100×).

locally obliterates the prevailing surface texture. Scuffing is usually considered a system failure. In an internal combustion engine, the cylinder just gets shiny with age if the lubricant is effective. If it is not, scuffing can occur and it is considered to be a wear failure.

Seizure

Seizure is adhesive wear at its worst. The rubbing parts "stick" together with sufficient bond strength that the mechanism no longer functions (Figure 3.9). As mentioned in the galling discussion, the formidable damage from galling is often seizure—the galling excrescence used up the running clearance.

Metal-to-Metal Wear

If a metal-to-metal sliding couple does not gall or seize due to adhesion, it may simply wear; small adhesions create wear particles and the particles stay between the rubbing

FIGURE 3.7
Scoring of a flat babbitt bearing in a lubricated sliding test versus cast iron.

FIGURE 3.8
Scuffing (scoring) on a hard steel punch produced by contact with a hard steel die (100×).

surfaces. They usually fracture into smaller particles, and if the particles start to separate the mating members, they lubricate to some degree and reduce the system wear.

Most metal-to-metal couples form a tribofilm under moderate loads and speeds (Figure 3.10). This tribofilm can contain wear debris, embedded lubricant, etc. It separates the rubbing surfaces and usually reduces system wear. With soft metals, the outermost surface usually contains debris and reaction products like oxides from the rubbing. Sometimes a nanocrystalline layer forms under this layer wherein the various materials form grains smaller than 100 nm in average size. A mechanically mixed layer can form under these two layers. This layer contains a mechanical mixture of the two mating metals. For example, if you rub a white crayon over an area covered with black crayon, you will eventually get a gray color. The colored waxes are mechanically mixed. Thus, from

FIGURE 3.9 (See color insert.)
Balls welded together by adhesive interaction (seizure) in a four-ball wear test for oils. One ball rubs on the lower three, which are held stationary, and the load on the upper ball is incrementally increased until the films separating the balls no longer separate and adhesive interaction (welding) occurs.

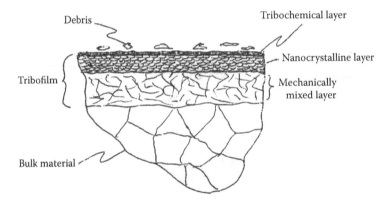

FIGURE 3.10
Cross section of a soft metal surface after sliding on another soft metal surface.

the microscope standpoint, the cross section of a metal-to-metal wear scar will look like Figure 3.10. Soft metals and atomically similar couples are very prone to these transformations. Hard metals are less prone. Usually these types of systems must be lubricated if sliding is more than infrequent/intermittent. In plastic sliding systems, one or both members can be filled with a PTFE, silicone, or other additive to mitigate adhesion. Figure 3.11 shows PTFE-bearing materials after a block-on-ring wear test (ASTM G 77). Figure 3.12 shows a lubricated plastic-metal couple after a very long service life. Wear on both members was very low. Lubricated plastic steel couples are widely used, and if the system is working properly, both surfaces simply polish, no scoring, no severe wear. The worn appearance of metal-to-metal sliding couples varies from polishing for well-lubricated hard-hard couples to simply lowering of surface texture height parameters (Figure 3.13). Oxidative wear is the usual result for unlubricated hard-hard couples (Figure 3.14) Both surfaces simply have the prevailing surface texture reduced in height (Ra, Rz, etc.) with continued sliding. Hard-soft couples are very common, as in lubricated powdered metal (PM) bronze bearings sliding on hardened steel shafts. Sometimes the bronze will transfer to the shaft and score marks

FIGURE 3.11
Appearance of plastic-to-steel wear. Two candidate PTFE fabric-bearing linings were rubbed against hardened steel rings in a block-on-ring test (ASTM G 77). The rings did not wear.

FIGURE 3.12
Appearance of a plastic-to-metal tribocouple after hundreds of millions of reciprocating cycles. The plastic wear is nil.

will be visible in the bronze and even in the steel. Some bronzes have hard second phases, and the adhesive wear can turn into abrasion when the bronze matrix gets eroded and the hard phases stand proud (Figure 3.15). If the soft member is very soft, as in lead-tin babbitt, the babbitt will usually score or scuff immediately after a lubricant film is diminished (Figure 3.16). For this reason, large babbitt bearings are often flooded with oil by operators on start-ups and shutdowns—there is insufficient shaft velocity to develop hydrodynamic film separation of the spaces. Soft metals on soft metals are used in many tribosystems (like brass mechanisms in clocks), but both wear severely (Figure 3.17) unless the loads are very low or if the surfaces are separated with a lubricant. Many antique clocks have been running for centuries with brass components, but they may not be the soft leaded brasses that are commonly mated in today's mechanisms. However, soft-soft couples can survive sliding speeds and loads if they are low enough (<100 PV).

FIGURE 3.13 (See color insert.)
Polishing wear of hard steel saw components after years of service.

FIGURE 3.14
Oxidative wear of a hard (60 HRC) ball rider (upper photo) on a counterface of equal hardness (ASTM G 133 reciprocating test). Both photos are magnified 100 times.

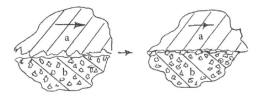

FIGURE 3.15
Adhesive wear progresses into abrasive wear as hard phases in (b) start to stand proud from the surface and abrade surface (a).

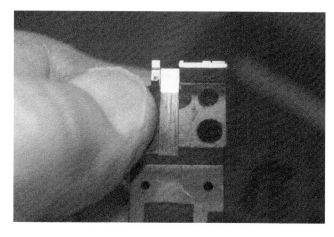

FIGURE 3.16
Scoring of a P/M bronze bearing after reciprocating sliding against a hard steel (60 HRC).

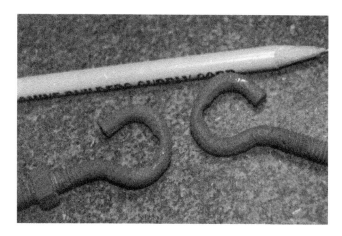

FIGURE 3.17
Wear of soft steel hooks that held a sign that was free to swing in the wind.

PV Limit

For decades in the United States and other countries, a PV number has been used to rank the efficacy of a plastic-metal couple. The concept is that there is a limiting load (P) and speed (V) for every plastic, below which it will work successfully in continuous sliding contact.

The English units for PV are:

P = apparent contact pressure in psi (MPa in SI)

V = rubbing velocity in ft/min (m/s in SI)

The limiting PV for a plastic bearing was obtained by lab testing, usually using a thrust bearing type of test specimen (Figure 3.18). Experimentally, the test speed would be held constant at a preselected value and the apparent pressure of the rubbing contact would be increased until high wear, velocity, or some other "failure" occurred. This would produce a limiting P at that test velocity. Then another velocity would be selected and the increased pressure sequence would be repeated. This testing is continued until a PV limit diagram is obtained (Figure 3.19). The plastic manufacturer would publish a PV limit for its plastic based upon these data. It also used these data to arrive at a specific wear rate for the plastic, which was essentially the k in the Archard equation. The specific wear rate could be used to calculate radial wear of a plain bearing. The use of PV data has waned with the consolidation of plastic manufacturers and the dearth of "new plastics," but it was, and still is, a useful tool in selecting a plastic for a bearing application. The obvious limitation of the concept is that the k values only apply to the PV conditions used in their development. For example, if a plastic had a PV of 10,000, users might think that they can run the plastic at 10,000 sfm (surface feed per minute) if the pressure is only 1 psi. Of course, it would melt. So PV data have to be applied only to those systems that operate in the PV range used in the PV testing.

In addition to staying within PV limits, care must be taken in using plastic plain bearings to accommodate their relatively high coefficient of thermal expansion—usually at least 10 times that of the surrounding metal housing. If sufficient clearance (usually 1% of shaft diameter) is not designed into the system, the bushing can expand heating and use up the running clearance. That is what happened on the bushing shown in Figure 3.20. The

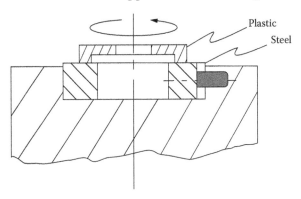

FIGURE 3.18
Typical plastic-to-metal wear test used to determine PV limits of plastics.

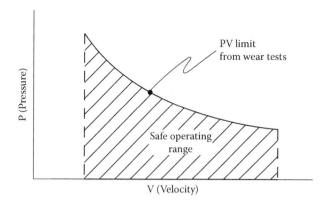

FIGURE 3.19
PV limit diagram for a plastic versus steel couple.

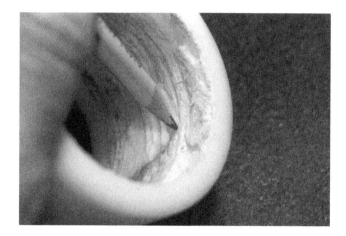

FIGURE 3.20 (See color insert.)
Bore of a plastic bushing after a temperature-induced seizure.

bushing was designed for low temperature service (below RT), and the failure occurred because the designer was not told that the machine was cleaned with 180°F water. Seizure occurred during the cleaning operation.

Soft-Soft Couples

One of the most common wear problems that involves adhesive wear is carbon graphite electric motor brushes sliding on soft copper motor commutators. The carbon graphite adhesively transfers to the copper, but the carbon graphite has enough nongraphite in it (ash, amorphous carbon) to abrade the copper and the transferred carbon graphite–ash also abrades the carbon graphite (Figure 3.21). As is the case in many wear systems, the wear process starts with adhesive wear, but may evolve into abrasion after the incubation adhesion stops due to separation by tribofilms.

FIGURE 3.21 (See color insert.)
Adhesive transfer of carbon graphite to a copper commutator from the carbon graphite brushes that rub on it.

Fretting Wear and Corrosion

Fretting was defined in Chapter 2 as small amphilate relative motion between contacting surfaces. In metals, the process (even well lubricated) starts by adhesion between the contacting surfaces (Figure 3.22). The starting surface finish is completely obliterated, and there are random grooves and pits and accumulations of wear reaction product (oxides for steels). The pits are a significant part of this wear process since they can be stress concentrations that lead to fatigue failures. Fretting wear is a fatigue process. Adhesive transfer occurs, the bonded material then gets stressed by the fretting motion, and it may fracture off, producing debris, or it may get rebonded to the other member and the process repeats. When particles fracture, the nascent surfaces react with the environment and fretting corrosion occurs. Figure 3.23 shows adhesive transfer from a hard steel surface to a hard steel rubbing counterface in the ASTM G 204 fretting corrosion test. Fretting wear is usually the process that occurs in plastics because there is no corrosion in the fracture process. The plastic particles do not oxidize on fracture like metals do in air. Fretting damage in transparent plastics usually appears as a "frosty" area (Figure 3.24). Fretting damage of plastics is very common, especially in packaging. Trade dress on cans/bottles, etc., may become obliterated from fretting motion in transport. It is often ignored unless it creates a frosty

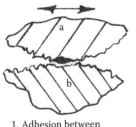

1. Adhesion between
a and b

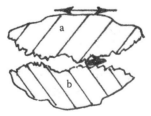

2. The adhered material
fractures to form
a particle

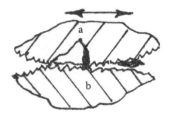

3. The process repeats
and pits form

FIGURE 3.22
The origin of fretting pits.

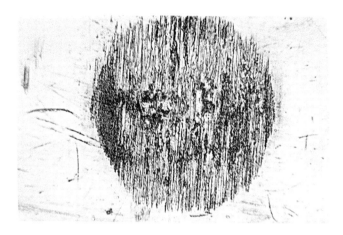

FIGURE 3.23
Adhesive transfer of hard steel on a hard steel counterface after a million fretting cycles (100×).

spot on an $80 pair of ski goggles or damage from rubbing on something in luggage or a backpack. Figure 3.25 shows fretting damage in a plastic-plastic couple (nylon rider, acrylic counterface). Figure 3.26 shows fretting damage on a hard steel plastic mold core.

In summary, adhesive wear usually is the predominating mode of wear during the "incubation period" for all types of solid-on-solid sliding wear. As we have tried to demonstrate, abrasion and corrosion can be conjoint, but the damage starts with adhesion between solid surfaces. It is intuitive in nature. Whenever any solid forcibly slides on another solid, something will happen if the forces exceed some property limit of either rubbing surface, like the compressive yield strength or shear strength. Figure 3.27 reviews the mechanism of adhesive wear. It is common practice in wear modeling to blame asperity interactions for starting wear. They are insignificant since they are usually obliterated at the first static contact of the mating couple. It is interacting wavelets on surfaces that determine the true area of contact. The wave contact points are likely only 1/1000 of the apparent area of contact, possibly a millionth; thus, the stresses are extremely high in the real contact areas. They are high enough to produce plastic deformation. They are high enough to extrude away separating films and produce atomic contact between surfaces and thus bonding. And these contacts are the origin of adhesive wear. They exist in all solid-solid contacts.

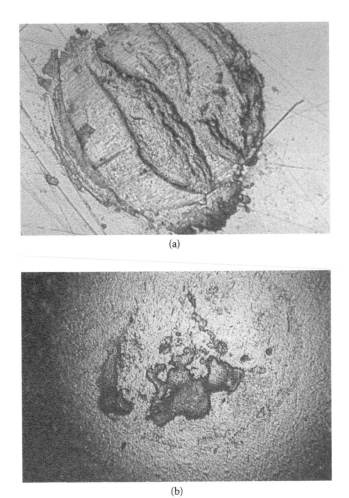

(a)

(b)

FIGURE 3.24
Fretting wear on an acrylic plastic flat (PMMA) (a) mated with a hard steel rider (b). The test surfaces were subjected to 10,000 rubbing cycles (100×).

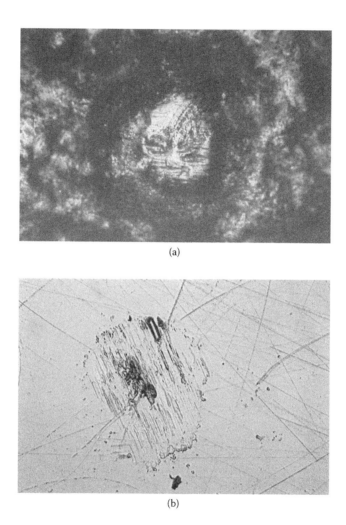

(a)

(b)

FIGURE 3.25
Fretting wear of a polyamide rider (a) on an acrylic flat (b) (100×). This was the same test as in Figure 3.24.

FIGURE 3.26
Fretting damage on a hard steel injection mold insert from clamping during mold closure.

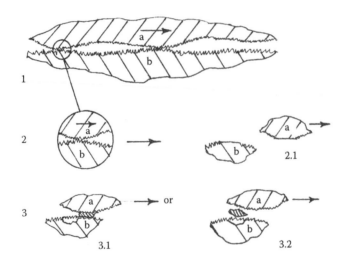

FIGURE 3.27
When surface (a) slides on surface (b) bonds can be formed in the real area of contact at interfering waves. The bonds can break (2.1), or material can fracture from one surface and adhere to the other (3.1), or adhered material can fracture and form wear particles (3.2).

Related Reading

ASTM G 77, *Standard Test Method for Ranking Resistance of Materials in Sliding Wear Using the Block-on-Ring Wear Test*, West Conshohocken, PA: ASTM International.

ASTM G 98, *Standard Test Method for Galling Resistance of Materials*, West Conshohocken, PA: ASTM International.

ASTM G 99, *Standard Test Method for Wear Testing with a Pin-on-Disk Apparatus*, West Conshohocken, PA: ASTM International.

ASTM G 133, *Standard Test Method for Linearly Reciprocating Ball-on-Flat Sliding Wear*, West Conshohocken, PA: ASTM International.

ASTM G 137, *Standard Test Method for Ranking Resistance of Plastic Materials to Sliding Wear Using a Block-on-Ring Configuration*, West Conshohocken, PA: ASTM International.

ASTM G 176, Standard *Test Method for Ranking Resistance of Plastics to Sliding Wear Using Block-on-Ring Configuration—Cumulative Wear Method*, West Conshohocken, PA: ASTM International.

ASTM G 196, *Standard Test Method for Galling Resistance of Material Couples*, West Conshohocken, PA: ASTM International.

ASTM G 204, *Standard Test Method for Damage to Contacting Surfaces Under Fretting Conditions*, West Conshohocken, PA: ASTM International.

Bikerman, J.J., *The Science of Adhesive Joints*, New York: Academic Press, 1961.

Buckley, D.H., *Surface Effects in Adhesion, Friction Wear and Lubrication*, Amsterdam: Elsevier, 1981.

Israelachvili, J.M., *Intermolecular and Surface Forces*, 2nd ed., San Francisco: Academic Press, 1992.

Johson, R.L., *Contact Mechanics*, London: Cambridge University Press, 1989.

Stachowiak, G., Batchelor A.W., *Engineering Tribology*, New York: Wiley, 2011.

4

Abrasion

Abrasion is the progressive loss of material from a solid surface caused by relative sliding in forcible contact with hard, sharp particles or protuberances. The protuberances statement means that files and saws produce abrasion, but the damage is not done with particles, but rather, hard, sharp cutting edges that stand proud. There are four modes of abrasion generally accepted by people working in tribology:

Low stress
High stress
Gouging
Polishing

These terms were defined in Chapter 2, and the different modes of abrasion have to do with the mechanics of the contact with the particles or protuberances that are causing the abrasion. The most commonly encountered abradants are components of the earth's crust: soil, rocks, mineral deposits, sand, etc. Most of these materials are inorganic and harder than many engineering materials like construction steels. Table 4.1 is the famous Mohs hardness scale that ranks the relative hardness of common minerals. It is a table of what scratches what. This chart shows that talc is the softest of the common minerals and diamond is the hardest, but also that sand (quartz) is harder than many minerals and is in most soils. Consequently, most digging and earth-moving activities involve contact with hard substances. Also, corundum, which is aluminum oxide by chemical formula, is a significant part of the earth's mantle. So, hard engineering materials frequently encounter minerals that tend to be harder than most materials, and therefore they are capable of causing abrasive wear.

Some tribologists include handling slurries (particles suspended in a fluid) as part of abrasive wear, but because slurries can be corrosive, slurry handling will be discussed in the chapter on tribocorrosion (Chapter 8).

Abrasion is everywhere. It is a part of the natural cycle of life. Very old buildings often have stone steps and floors that have lost materials to abrasion over the years (Figure 4.1). The abrasive substances are sand and the like rubbed on the surface by footwear. This kind of abrasion and the other kinds of abrasion that were mentioned can be mitigated; for example, harder entry materials, such as granite, could be used. A person who loses a favorite pair of shoes to sole and heel wear probably wishes that there was a more abrasion-resistant material for construction of them. There is an elastomer that will outwear leather and most rubbers by orders of magnitude, and these kinds of abrasion-resistant materials are identified by laboratory abrasion tests, some of which will be discussed later. This chapter will help readers identify modes of wear, present proposed wear mechanisms, enumerate abrasion manifestations, and show how abrasion differs from other wear and erosion processes.

TABLE 4.1

Mohs Hardness Scale

Material	Characteristics	Mohs Hardness
Talc	Barely scratched by fingernail	1
Gypsum	Scratched by fingernail	2
Calcite	Scratches copper and is scratched by it	3
Fluorite	Not scratched by copper; does not scratch glass	4
Apatite	Just scratches glass	5
Orthoclase	Just scratched by a file, easily scratches glass	6
Quartz	Not scratched by a file	7
Topaz	Scratches quartz	8
Corundum	Scratches topaz	9
Diamond	Scratches corundum	10

FIGURE 4.1 (See color insert.)
A stone entry step worn concave by centuries of abrasion from hard particles on footwear.

Mechanisms

The operative word in the definition of abrasion is *hard*. A material that is significantly harder than another can damage the softer in static or sliding contact. For example, just setting a 5 kg rock on a finely finished wood table will likely produce dents in the wood at points of contact. If the same rock is slid on the wood table, it will produce scratches (Figure 4.2). The scratches can be deformed grooves with material displaced from them, or the grooves can form by chip removal combined with furrow formation (Figure 4.3). If the rock is rolled on the finely finished wood table, it will produce dents and possible material removal from chip formation at points of contact (Figure 4.4). Material can also be removed from the softer surface by adhesive transfer to the harder surface (Figure 4.5). Studies on sand particles used in abrasion testing show that this mechanism is likely predominating

FIGURE 4.2 (See color insert.)
Scratching abrasion from sliding a rock on a finely finished wood surface. (Do not attempt replication of this experiment without permission of the owner of the finely finished surface.)

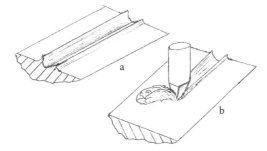

FIGURE 4.3
Scratching abrasion can form a furrow by plastic deformation (a); it can also produce a chip by fracture along with the furrow (b).

FIGURE 4.4 (See color insert.)
Indentation of a wood surface from rolling contact with a rock.

FIGURE 4.5
Rolling or sliding particles can remove material from a solid surface by adhesive transfer of the surface material to the abradant.

in applications involving rolling as opposed to sliding of hard substances on softer surfaces. This is adhesive wear, as was discussed in Chapter 3.

An important aspect of abrasion is the issue of rolling on the surface or sliding. It is common to categorize abrasion produced as two-body or three-body abrasion. Two-body abrasion is produced by fixed abrasives. Three-body abrasion is produced by abrasive particles that are forced against a soft surface by something, a third body. The third body could be a solid or simply compacted particles, as in tilling soil. The soil particles are forced against the tilling tool by untilled soil.

Figure 4.6 illustrates the concept of fixed abrasives. Approximately same size stones were "fixed" into a concrete cap on a stone wall to keep tourists from sitting and walking on the wall, but this is how fixed abrasive papers (sandpaper) and films are made. Approximately the same size abrasive particles are bonded to a flexible substrate with a thin film of adhesive. Grinding and cutoff wheels are made by adhesive bonding particles together (resin bonded wheels) or by using a glass as the glue between particles (vitrified wheels). Fixed abrasives tend to produce scratching abrasion (Figure 4.3), while loose abrasives can roll to indent a soft surface or become momentarily fixed to produce scratching abrasion. The concept of two-body versus three-body abrasion is illustrated in Figure 4.7.

Both two-body and three-body abrasion remove material, but two-body is usually more effective. Figure 4.8 shows scratches produced by the same size abrasive particles in fixed mode (two-body) and loose mode (three-body). The scratches are much deeper from the

FIGURE 4.6
Fixed abrasives are manufactured like this wall with embedded stones (to deter people from sitting or walking on it). The abradants can be affixed to a flexible media like paper (sandpaper), or they can protrude from a rigid body as in a grinding wheel.

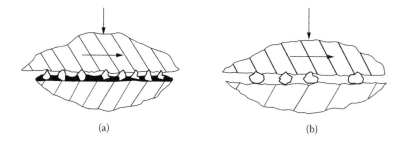

(a)   (b)

FIGURE 4.7
Two-body abrasion is shown in (a); the abrasive particles are fixed to a substrate during rubbing. In (b), the particles are forced against the fixed surface by the third body.

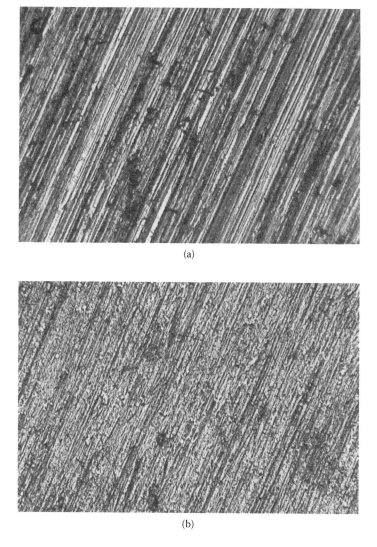

(a)

(b)

FIGURE 4.8
Three-body abrasion of 1020 steel by 60-grit silica; 0.25 m rub, 55 N force (a); (b) shows two-body abrasion under the same sliding conditions. Both photos are magnified 100 times.

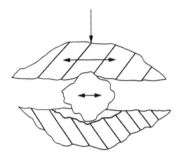

FIGURE 4.9
Pitting produced by loose abrasive particles not moving far in lapping and other three-body abrasion processes.

fixed abrasive than the loose because the loose abrasive was forced against the softer surface by a resilient material: rubber. This is low-stress abrasion.

Polishing often occurs in handling loose hard substances. If the forces pressing the abrasive on the softer surface are low, the adhesion mechanism of material removal will likely prevail. If the forces are high, scratching combined with adhesion will probably occur. When the forces get high enough to crush the abradant (high-stress abrasion), scratching and possibly pitting will be the mechanisms for material removal. Rolling grits can pit when, for example, a grit does not roll far, but sort of rubs back and forth in place. It "burrows" itself into the softer material to produce a pit (Figure 4.9). In scratching abrasion only a small percentage of a particle's shape penetrates to form a scratch; maybe only 1% of the particle diameter penetrates and slides to form a scratch. On the other hand, if a loose abrasive particle is subjected to sliding conditions that allow it to stay in one spot, it, or groups of like particles, can form pits. This is commonly encountered in lapping operations that allow particles to stay in place.

Polishing can occur with fixed abrasives if the abrasive is fine enough to reduce the prevailing surface texture on the softer surface. Intentional polishing is usually accomplished by embedding loose abrasive particles in a compliant medium and rubbing them on the surface to be polished by the motion of the compliant medium (Figure 4.10). The mechanism of material removal is usually adhesion of material from the softer surface to the harder particles.

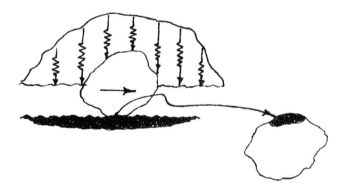

FIGURE 4.10
Schematic of polishing particles held by a compliant medium.

Gouging abrasion usually involves material removal by multiple indentations—a dent, redented, and redented again until a piece of material fractures from the softer surface. This is conjoint with scratching where material is removed by rubbing.

In summary, the mechanism of material removal in abrasion depends on the mode of abrasion, but scratching and adhesive transfer to abradants are the predominant mechanisms for material removal.

Manifestations

Low-Stress Abrasion

Tools that move earth lose their shape and take on a dull appearance (Figure 4.11). They encounter mostly three-body abrasion from sand and similar hard particles in the earth. The scratching that occurs under low-stress abrasion depends on the size of the abrasive particles encountered and the force that presses the particles against a surface (Figure 4.12). In the case of abrasion by protuberances, the important factor is the size of the protuberances. Fine files produce small scratches; course files produce large scratches. Fixed abrasives produce scratching abrasion with the scratch depth a function of the abrasive particles in the grinding wheel or abrasive paper/film. Figure 4.13 shows scratches produced in a hard steel surface after surface grinding with a 60-grit grinding wheel. This is low-stress abrasion; grit does not fracture under low-downfeed conditions. Low-stress abrasion on a mason's trowel is shown in Figure 4.14, and low-stress abrasion from asbestos packing on stainless steel looks like Figure 4.15. Finally, low-stress abrasion from rolling grit in lapping mode is shown in Figure 4.16. Scratching abrasion is replaced mostly by adhesive removal of material and pitting.

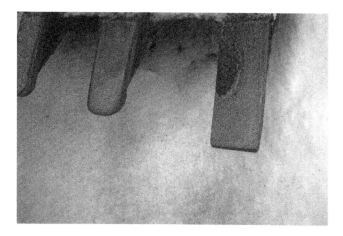

FIGURE 4.11
Abrasion of a backhoe bucket tooth from digging in sandy soil. Note the shape change from new condition (right).

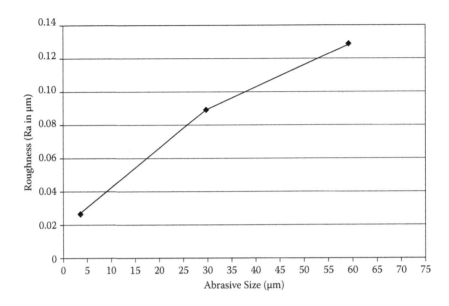

FIGURE 4.12
The surface roughness generated by various size fixed abrasives in a single pass at a fixed force (on type 430 stainless steel).

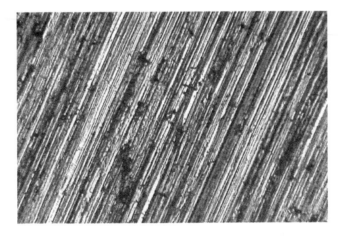

FIGURE 4.13
Scratches produced by a 60-grit alumina wheel on hard steel (60 HRC). The magnification is 100 times.

Abrasion of Plastics and Other Non-Metals

Plastics scratch and pit much like metal under low-stress abrasion conditions. They usually display lower abrasion resistance than soft steels in laboratory sand abrasion tests, but there are many applications where they are still used. Plastic snow shovels are common, and they abrade from rubbing on rough concrete (Figure 4.17). Plastic ski boots abrade from walking on snow-free concrete and roadway surfaces (Figure 4.18). The more rigid plastics scratch like metals (Figure 4.19). The lower elastic modulus plastics tend to produce fibrils when abraded (Figure 4.20). These are shreds of plastic plowed from furrows, but still affixed to the worn member. The softer the plastic, the greater the tendency to form attached abrasion fibrils.

(a)

(b)

FIGURE 4.14 (See color insert.)
Scratches at 100× produced by sand (a) and aggregate on the hard steel of a mason's trowel (b).

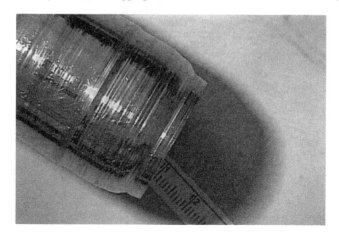

FIGURE 4.15 (See color insert.)
Abrasion produced by asbestos packing on a stainless steel bushing.

(a)

(b)

FIGURE 4.16
(a) Abrasion produced by lapping soft steel (a) with 100 µm alumina; (b) shows the resulting surface texture at 100×.

FIGURE 4.17 (See color insert.)
Abrasion of a plastic ultrahigh molecular weight polyethylene (UHMWPE) wear strip on a plastic snow shovel.

FIGURE 4.18 (See color insert.)
Abrasion of plastic heels on a ski boot from walking in parking lots free of snow.

Rubbers tend to form roller and fracture lines resembling tree bark (perpendicular to the grinding direction when they are ground with fixed abrasives). The surface texture of abraded rubber, however, depends on the type of rubber and its hardness. A 60 Shore A neoprene has a matte "pitted" appearance when abraded by a 30 μm finishing tape. Some rubbers like styrene butadiene rubber (SBR) tend to have a very smooth, almost shiny, surface under three-body abrasion conditions. Most auto tires are SBR, and they often wear smooth (Figure 4.21).

Ceramics tend to abrade by scratching abrasion with clean furrows and small furrow lips (Figure 4.22). They are brittle and harder than most substances that abrade, so abrasion usually requires action from hard particles like aluminum oxide. Many ceramics can be ground with aluminum oxide grinding wheels. However, alumina ceramics require diamond. Our hardest ceramic, diamond abrades by the adhesion component of abrasion. For example, single-crystal diamonds used to cut glass in manufacturing operations readily "abrade" by the glass. The glass does not scratch the diamond, but carries away diamond possibly at the carbon atom level. Carbon adheres to the silica.

In summary, low-stress abrasion can have different visual appearances and different appearances under the microscope. Scratching abrasion almost always occurs when fixed abrasives are encountered. Loose abrasives can produce scratches as well as pitting. The pitting tends to occur when the abrasive particles are large and the forces pressing them on a surface are large and the motion is restricted. Microscopic examination of metals (maybe only 10×) will tell a user which type of material removal predominated on worn parts that are being analyzed. It is almost always helpful to consider what is causing the abrasion and how it removes material from the surface.

High-Stress Abrasion

Figure 4.23 shows what high-stress abrasion looks like on a hard steel brick hammer. The appearance can vary from scratched (hard steel) to pitted from rolling grit on soft steel, to almost polished on cemented carbide. It looks similar on a digging tool used on rocks. Fortunately, bulldozer and power shovel teeth do not have to be very sharp, but they lose efficiency as the edges get blunter (Figure 4.11). Ceramics are not usually used for applications involving high-stress abrasion because they are prone to brittle fracture

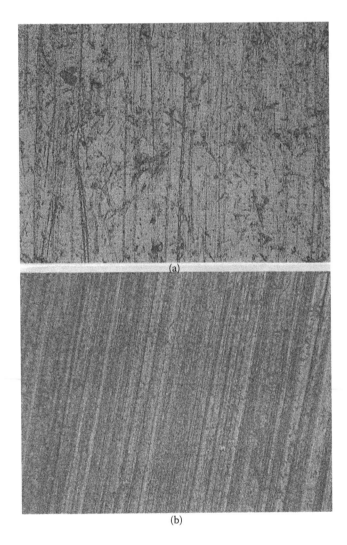

FIGURE 4.19
Abrasion of an acrylic plastic (PMMA) by 60-grit silica under three-body conditions (a) and two-body conditions (b). Both photos are at 100 times magnification.

just like rocks and mineral shapes. They scratch when ground with fixed abrasive wheels (Figure 4.22), and they tend to smooth under the action of loose abrasives.

Gouging Abrasion

Gouging abrasion produces a surface mostly consisting of overlapping indentations (Figure 4.24) caused by plastic deformation of the surface. As mentioned previously, tough steels or steels that work harden are commonly used for gouging abrasion applications. Spike heels on a wood floor will cause gouging, but usually there is no material loss, so this damage is not really gouging abrasion. Truck bodies that carry rocks usually show gouging abrasion, as do rock-crushing machines.

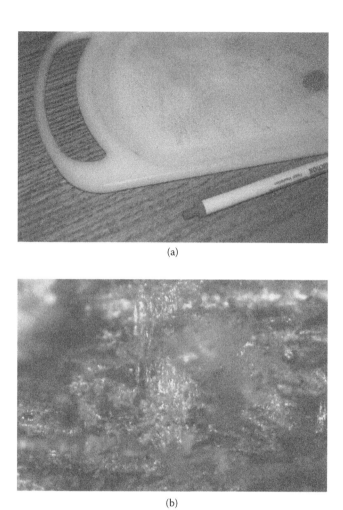

(a)

(b)

FIGURE 4.20
(a) Polypropylene plastic cutting board. (b) Abrasion of board (100×).

Polishing Abrasion

There is not a universally agreed to definition of a polished surface, but most will agree that it is a surface with a specular reflectance. Images can be mirrored on it. A mirror surface usually has a surface roughness less than 0.025 µm. A surface roughness of 0.1 µm is shiny, but will not act as a mirror—scratches from finishing will be visible with a 10× loupe. Since a good polish has a roughness less than 0.025 µm Ra, the abradant that generated the polished surface has to be small—usually less than 1 µm in diameter. So polishing abrasion commonly is produced by hard particle substances that are less than 1 µm in diameter. Figure 4.15 shows uneven polishing of a stainless shaft from rubbing versus asbestos packing. Polished surfaces on most materials appear featureless under the microscope, except for defects like pores and inclusions, and multiphase materials show the phases present if they respond differently to polishing (Figure 4.25).

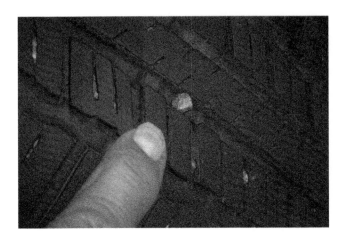

FIGURE 4.21
Abraded rubber on an automobile tire.

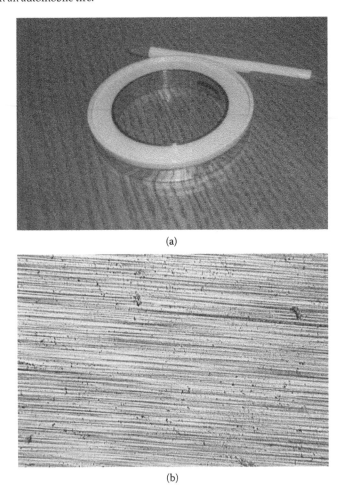

FIGURE 4.22
(a) The rubbing face of a zirconia seal. (b) The ground surface is slightly polished from the rubbing action (100×).

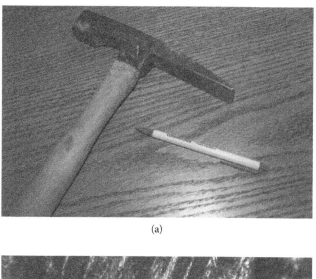

(a)

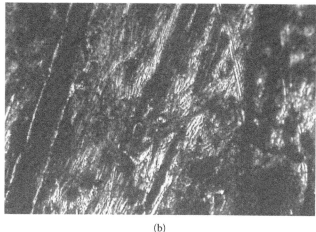

(b)

FIGURE 4.23
The appearance of high-stress abrasion on the cutting edge of a stone mason's hammer (100×). (a) Stonemason's hammer. (b) The appearance of high-stress abrasion on the cutting edge (100×).

Abradants

Abrasion by the currently acceptable definition requires a hard substance acting on a softer substance. Figure 4.26 shows the category of materials that are harder than steels, but abrasion affects all solids, so relative hardnesses of nonferrous materials, plastics, etc., need to be given consideration when the question arises: What solid is harder? Table 4.1 shows the Moh's hardness scale for materials. It shows what scratches what. Unfortunately, important engineering materials are not on the scale; it does not tell us which of the minerals will scratch soft steel. Figure 4.26 is an attempt to compare the hardness of engineering materials with some minerals. The point is that when solids of differing hardnesses rub together, the harder usually becomes the abradant, and it damages the softer material, while it may not be damaged or sustain lesser damage or be the exception of adhered par-

FIGURE 4.24
Gouging abrasion on the "cutting" wheels on a trash grinder. The wheels are 60 mm thick.

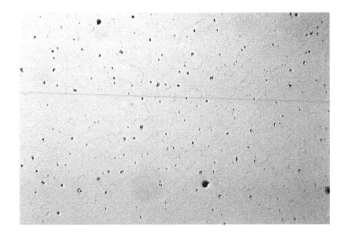

FIGURE 4.25
Metallographic polished stainless steel (100×). The surface is scratch-free at this magnification, but the pores in the steel are visible.

ticle from the softer material. Besides relative hardness, some other factors that affect the ability of a material to abrade other materials are:

- Friability/strength
- Particle shape/size
- Elastic modulus
- Fixed or loose

Friability is the ease with which an abrasive particle fractures under load. Silicon carbide is widely used for fine sandpapers, but anybody who has ever used these papers for big projects knows that they do not last long. The grit fractures and the ability to abrade diminishes greatly. It is friable, while aluminum oxide is not. The particles are stronger and less likely to fracture when used as an abradant.

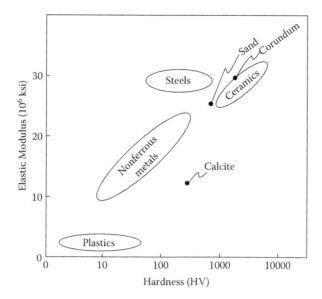

FIGURE 4.26
The hardness of abrasive substances (sand, alumina, and calcite) compared with the hardnesses of engineering materials.

Particle shape and size are very important considerations when it comes to an abradant's ability to remove material. Blocky particles tend to roll on surfaces, while more angular abrasives will be more likely to produce scratching abrasion that generates chips. Figure 4.27 illustrates the common shapes taken on by abrasive particles in nature. Abradants produced by fracture processes tend to be angular. Probably more important than shape is size. Larger abradants do most of the damage (abrasion). In many applications where the abrasive substance is trapped between two solids, only the larger particles carry the load. For this reason, particle abradants need to be analyzed for their particle size distribution to be ranked for ability to abrade.

FIGURE 4.27
Abradant particles can have smooth or rough surfaces, and their shapes can vary from round to angular to blocky to sliver shaped.

The elastic modulus of an abradant goes hand in hand with hardness in determining an abradant's ability to abrade—the higher the abradant's elastic modulus, the greater its ability to abrade. For example, steels have a higher elastic modulus than any nonferrous material or plastic. The softest steel will scratch the hardest plastic. The hardest plastic will probably have an elastic modulus of 10^6 psi (7 GPa) compared to steel's 30×10^6 psi (207 GPa).

The final factor on abrasion is fixed or loose particles, and this is extremely important in determining the ability of an abradant to remove material from a softer surface. Fixed abrasives are better abradants than loose abrasives. Tilling soils and mining usually involve contact with loose abrasives, while intentional abrasion (grinding) is usually performed by fixed abrasives. Fixed abrasives are thought by many to dull and thus become ineffective. More often than dulling, loading with wear debris and adhered material make fixed abrasives less abrasive. In any case, the important point to remember is that fixed abrasives are better abradants than loose. They are more efficient in wearing material because sliding is more likely to produce a scratch and adhered material than a loose rubbing particle.

Abrasivity

Unfortunately, the factors (hardness, size, shape, etc.) that control a material's ability to abrade other materials, its abrasivity, are not readily available for the soils and rocks that are encountered in agriculture and mining. The same situation exists for process fluids, like drilling mud. The abrasivity of a particulate substance can be measured by comparing it to materials with known abrasivity. For example, to determine the relative abrasivity of dust from various parts of the moon, the dust was rubbed on steels and other materials, and the damage it produced was compared with the damage produced by silica and alumina under the same rubbing conditions. Relative abrasivity can be measured in a laboratory bench test with a rig like that shown in Figure 4.28. A rubber wheel continually rubs the particulate material against test specimens, and the mass loss is the test metric. The abrasivity of a particulate substance can be measured.

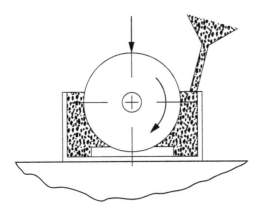

FIGURE 4.28
Schematic of a test rig that uses a rubber wheel immersed in particles to assess the abrasivity of those particles on a flat, horizontal test specimen.

Measuring Abrasion Resistance of Engineering Materials

The most widely used wear test (based on literature citations) is the ASTM G 65 dry sand rubber wheel abrasion test. This test is shown schematically in Figure 4.29. A test specimen is forced against a rotating rubber wheel and silica sand (50 to 70 mesh) is metered in a flat stream into the rubbing contact. A crescent-shaped scar is made in the test specimen (Figure 4.29), and the test metric is wear volume obtained from mass change measurements. The scar appearance varies from scratches for softer materials to an almost polished surface for hard materials. The ASTM standard test document supplies the user with data on reference materials of different hardnesses, and this allows a user to assess his or her test materials—the abrasion resistance compared to known materials used for abrasion applications like type D2 tool steel. This is a low-stress abrasion test because the rubber wheel limits the force, tending to crush the abrasive such that the abrasive is thought not to fracture in traveling through the contact.

If an application involves stresses that are likely to fracture an abradant, the ASTM B 611 test produces high-stress abrasion. It is similar in concept to the G 65 test except the wheel is low-carbon steel, the abrasive is aluminum oxide instead of sand, the particle size is about 1000 µm versus 300 µm, and the abradant is introduced into the contact in an agitated water slurry. The force of the wheel on the test specimen is sufficient to crush the alumina abrasive. This test is not intended for soft materials. In fact, it was established as

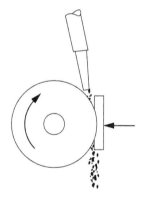

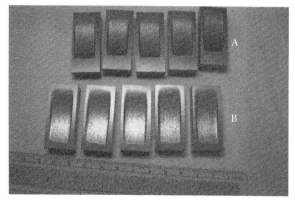

FIGURE 4.29
Schematic of the ASTM G 65 sand abrasion test and typical test specimen results.

a ranking test for the abrasion resistance of cemented carbides. However, it has been successfully applied to high-hardness weld hardfacings to ceramics and cermets. The wear scars looks similar to those produced by the ASTM G 65 test. Corrosion is not thought to be a significant contributor to the material loss in this abrasive test because the test time is 10 or 20 min and the corrodent is only distilled water.

The ball cratering or microabrasion test is very popular in Europe for ranking abrasion resistance of materials. It is shown schematically in Figure 4.30. A steel ball about 25 mm in diameter rotates on a flat specimen and a water slurry of fine abrasives (3 to 10 µm) is metered into the contact. Some rigs just use the weight of the steel ball as the load; others can apply significant loads—instead of a ball, a shaft with a ball shape in the middle is mechanically forced against the test specimen. The test metric is wear volume, and since a spherical crater is produced, the wear volume can be calculated as a sector of a sphere. The usual abrasive is 5 µm silicon carbide, but others can be used.

If an application is better simulated by fixed abrasives, the ASTM G 174 test may be applicable. This test, which is illustrated schematically in Figure 4.31, uses a loop of abrasive finishing tape as the abradant. The test specimen is a small flat that is tangentially loaded against the tape at the spindle that drives the tape. The standard has various procedures for soft steels to cemented carbides. The abrasive loops are aluminum oxide in 3 or 30 µm diameter particles. Wear volume is the test metric after a given number of loop passes. This test can rank all materials to low-stress abrasion by one of the most aggressive abradants, aluminum oxide. The test is very good for thin coatings, such as physical vapor

FIGURE 4.30
Microabrasion test (ball cratering).

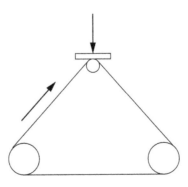

FIGURE 4.31
Schematic of the ASTM G 174 loop abrasion test. The test specimen is abraded by a fixed abrasive loop.

FIGURE 4.32
ASTM G 174 test specimens after testing.

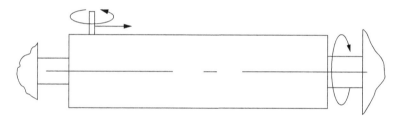

FIGURE 4.33
Schematic of the ASTM G 132 pin abrasion test.

deposition (PVD) and chemical vapor deposition (CVD), diamond-like carbon (DLC), and the like. G 174 wear scars are shown in Figure 4.32. A more aggressive fixed abrasive test is the ASTM G 132 pin-on-sandpaper test (Figure 4.33). Any fixed abrasive is applied to a large drum. A revolving pin on the test material is loaded on the drum and traversed along the drum such that it always sees fresh abrasive. The test metric is wear volume. The abradant can be any abrasive that is available in a format that will allow it to be applied to a drum.

In summary, there are bench tests that are widely used to rank the relative abrasion resistance of materials intended to resist abrasive wear in various applications. A user can opt for a high-stress or low-stress test, loose abrasive, or fixed abrasive. There is even a gouging abrasion test (ASTM G 81), so it is possible to combat abrasive wear using laboratory rankings of candidate materials.

Differentiation of Abrasion from Other Modes of Wear and Erosion

It is common for beginning tribological researchers to spend considerable time, money, and effort in analyzing wear scars to determine if abrasion or some other mode of wear occurred in a bench test. If the wear scar shows scratches, they conclude that abrasion is the mechanism of material removal. However, metal-to-metal wear with no hard particles present can produce the same appearance. Metal-to-metal wear produces "mechanical mixing" of the surfaces if the surfaces are capable of plastic deformation, and most metals are unless they are harder than 60 HRC or coated/treated. The surface of one metal can adhere to the other; then the adhered material can transfer back to the original surface and the process repeats until both surfaces interpenetrate in a mechanically mixed layer, a tribolayer. These tribolayers can contain protuberances of work-hardened material that can scratch the other surface, and thus scratching can occur with no apparent abradants in the tribosystem. Therefore, the factor that makes abrasive wear different from adhesive wear is the presence of abrasive particles/substances other than those that are products of the wear process itself. Abrasive wear scars can have an appearance from gouging to scratching to mirror finish polishing and everything in between. Abrasion requires a harder substance imposed on a softer substance, and the harder substance is usually from the mineral family, usually inorganic material, very often an oxide. Most elements occur in nature in the form of a chemical compound, and these compounds tend to be rock or rock-like or soil-like in nature.

Abrasion occurs in lubricated sliding systems when hard substances become contaminated in oils or greases. The classic example is diesel soot in the oil of diesel engines. It has a deleterious effect on wear life of engines, as is well documented. Dust particles in oil turn it into a lapping compound. Even nanoparticles can exacerbate lubricated wear. The nuances of abrasion existing in metal-to-metal or other "clean" tribosystems are not significant compared to the abrasion that occurs in tribosystems that operate in known abrasive substances: soil, rocks, minerals, and the like. The huge costs of abrasion are in vehicle tires, tilling tools, mining tools, drilling tools, earth-moving tools, comminution tools, and the like—machinery exposed to substances that are known to be harder than many metals.

All engineering materials abrade by scratching, pitting, or adhesion transfer to abrading substances; the damage can look like polishing, scratching, or gouging, and the most common remedy is to make the abraded surface harder than the abrading substance. This is often not that easy to accomplish.

Related Reading

Adamiah, M., Ed., *Abrasion Resistance of Materials*, Intech Co., 2012.

ASTM B 611, *Standard Test Method for Abrasive Wear Resistance of Cemented Carbides*, West Conshohocken, PA: ASTM International.

ASTM G 56, *Standard Test Method for Abrasiveness of Ink Impregnated Fabric Printer Ribbons and Other Web Materials*, West Conshohocken, PA: ASTM International.

ASTM G 65, *Standard Test Method for Measuring Abrasion Using the Dry-Sand Rubber Wheel Apparatus*, West Conshohocken, PA: ASTM International.

ASTM G 81, *Standard Test Method for Jaw Crusher Gouging Abrasion*, West Conshohocken, PA: ASTM International.

ASTM G 132, *Standard Test Method for Pin Abrasion Testing*, West Conshohocken, PA: ASTM International.

ASTM G 171, *Standard Test Method for Scratch Hardness of Materials Using a Diamond Stylus*, West Conshohocken, PA: ASTM International.

ASTM G 174, *Standard Test Method for Measuring Abrasion Resistance of Materials by Abrasive Loop Contact*, West Conshohocken, PA: ASTM International.

ASTM G 195, *Standard Test Method for Conducting Wear Tests Using a Rotating Platform Double-Head Abraser (Taber Abraser)*, West Conshohocken, PA: ASTM International.

James, D.I., Jolley, M.E., *Abrasion of Rubber*, New York: Maclaren, 1967.

Liang, H., *Tribology in Chemical-Mechanical Planarization*, Boca Raton, FL: CRC Press, 2005.

Marinescu, I.D., Rowe, W.B., Dimitrou, B., Inasak, I., *Tribology of Abrasive Machining Processes*, Norwich, NY: William Anderson Publishing, 2004.

Totten, G.E., *Mechanical Tribology: Materials Characterization and Applications*, Boca Raton, FL: CRC Press, 2004.

5

Rolling Contact Fatigue

Rolling contact, as opposed to sliding contact, predominantly occurs in countless industrial devices:

Tires on pavement

Wheels on tracks

Balls in raceways

Rollers in raceways

Gear teeth (some)

Rollers in conveying

Metalworking rollers

Wheels on shopping and utility carts

The first two may be the most costly problems. Vehicle tires fatigue roadway pavement (Figure 5.1), resulting in billions of dollars of roadway repairs. Fatigue on railway tracks (Figure 5.2) is similarly costly when the damage results in rail replacement. The next two can also be very costly when the balls or rollers are in bearings in a vehicle or piece of machinery. A simple spur gear "rolls" on its mating tooth for some portion of its engagement. Countless materials are moved on conveyor rollers and metalworking rollers supply us with most wrought metal products: sheet, plate, strip, structural shapes, etc. All of these applications are worldwide, and all can be very expensive when damage requires replacement.

What happens? What is rolling contact fatigue wear? Rolling contact fatigue is material removal/damage caused by repeated rolling of a solid shape on a contacting solid surface. The usual manifestations include pitting (Figure 5.3), micropitting (Figure 5.4), cracking (Figure 5.5), spalling (Figure 5.6), indentation (Figure 5.7), and brinelling (Figure 5.8). There are many nuances to these manifestations, but these examples present the "flavor" of this form of wear. This chapter will describe the general mechanism of rolling contact fatigue, and then describe in more detail micropitting, pitting, spalling, and slip in rolling contacts.

Mechanism

One more definition is in order. The ASTM G 2 Committee on Wear and Erosion defines *rolling* as "motion of a sphere, cylinder, or revolute shape in a direction on another surface characterized by no relative slip between contacting surfaces" (Figure 5.9). Thus, when a ball roller or wheel is loaded on another surface, the real area of contact is much different than the apparent area of contact. The term used for point or line contacts is *Hertzian contact*. This term acknowledges the work of Helmholtz Hertz in writing the equations for

FIGURE 5.1
Rolling contact fatigue of a macadam roadway.

FIGURE 5.2
Rolling contact fatigue on a railroad track that produced spalling.

FIGURE 5.3
Pitting on a frequently used tram track.

FIGURE 5.4
The frosted area on this roller (25 mm major diameter) was caused by micropitting fracture of microscopic particles from the rolling surface.

FIGURE 5.5 (See color insert.)
Cracking on a railroad track.

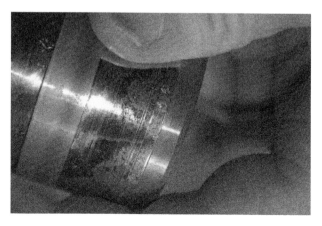

FIGURE 5.6 (See color insert.)
Spalling of a hardened steel rolling surface for cam followers.

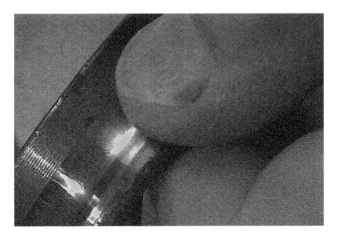

FIGURE 5.7 (See color insert.)
Indentation in a ball bearing raceway from contaminant particles.

FIGURE 5.8 (See color insert.)
Brinelling of a raceway from static overload.

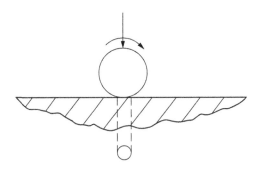

FIGURE 5.9
Pure rolling, no relative slip, only occurs in a ring the size of which depends on the force.

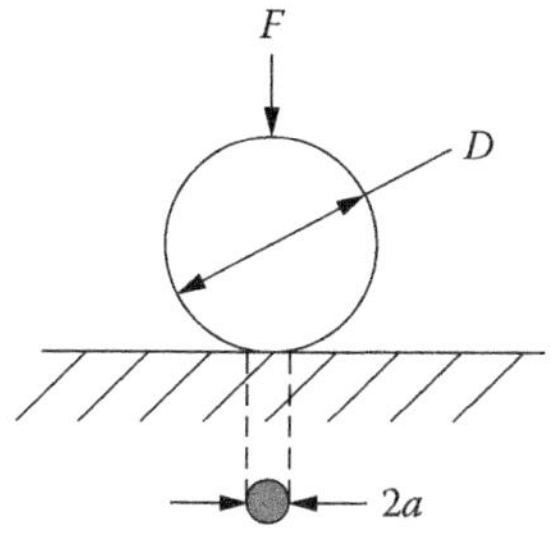

$$a = 0.721 \sqrt[3]{FD\left(\frac{1 - v_1^2}{E_1} + \frac{1 - v_2^2}{E_2}\right)}$$

Max stress, $S = \dfrac{1.5F}{\pi a^2}$

Where

a = radius of the contact area
D = the diameter of the sphere
E_1 = elastic modulus of the sphere material
v_1 = Poisson's ratio of the sphere material
E_2 = elastic modulus of the counterface
v_2 = Poisson's ratio of the counterface
F = force on the sphere

FIGURE 5.10
Hertz equation for the contact of a sphere on a flat surface under elastic conditions.

the stress in the contact region of elastic solids, with shapes such as spheres and cylinders contacting other spheres or cylinders or mated with flat surfaces. Figure 5.10 illustrates Hertz's formula for the real contact area between a ball and flat surface. There are formulas for most common geometric shapes, and finite element analysis can be used to calculate the contact stress on nontraditional shapes.

As can be seen from the equation in Figure 5.10, the real area of contact is a function of the size of the rolling shape (radius), the elastic modulus, and Poisson's ratio of both members, the force pressing the bodies together. The higher the elastic modulus, the smaller the contact area. For this reason, hard-on-hard couples, as occur in most rolling element bearings, can result in huge loads being carried by very small metal contacts. It has been known for a very long time that the highest stress in a material subject to a Hertzian contact is not on the surface of the material, but the subsurface (Figure 5.11). It is this subsurface stress that produces the tendency for rolling contacts to produce subsurface cracks that propagate to produce material removal (Figure 5.12). Spalling is defined as material removal from a solid surface in the form of a platelet or similar shape caused by propagation of subsurface fractures to a free surface. In the case of coatings, spalling usually produces a flake or platelet of coating.

In summary, the mechanism of rolling contact fatigue is fracture of material from a solid surface that is subject to repeated compressive stressing from contact with a rolling body. The pieces or particles that fracture from a rolling fatigue-affected surface can be microscopic or macroscopic, depending on the tribosystem. The material removal starts with the development of a subsurface crack that, with repeated stressing, propagates upward in multiple directions so that a piece of material is free to be ejected from the surface.

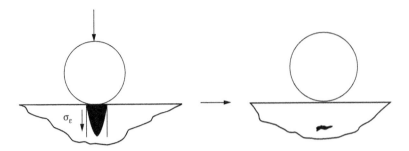

FIGURE 5.11
Peak stress in the compressive stress profile under a sphere on a flat surface.

FIGURE 5.12
Spalling of hard chromium plating on the hard steel shaft after reciprocating service with a ball bushing. Each ball produced a spall.

Micropitting

Micropitting is material removal by the formation of microscopic craters from surfaces subject to rolling fatigue conditions. Figure 5.13 shows incipient micropitting in a well-used ball bearing. Micropitting is common in ball and roller bearings. When a well-used rolling element bearing is disassembled and the rolling elements or raceways are examined with the unaided eye, a "line of travel" is often present. This is usually casually passed off as the real area of contact of the balls or rollers with the raceway. It is that, but it is not caused by the balls or rollers squashing down surface texture features. It is usually micropitting. Magnification of the line of travel shows that the frosted appearance is due to tiny pits where material spalled from the surface.

Subsurface cracks are promoted in bearing steels by nonmetallic inclusions such as hard phases (carbides), sulfides, oxides, etc. For this reason, bearing steel manufacturers go to great lengths to minimize inclusions and make those present as small as possible. Heterogeneous steels with large inclusions have poor resistance to rolling fatigue. Figure 5.14 shows the microstructure of two tool steels. One contains massive carbide particles, the other none. Micropitting patterns on raceways are often used to diagnose operating problems with bearing construction, lubrication, race shape, etc. If a rolling element

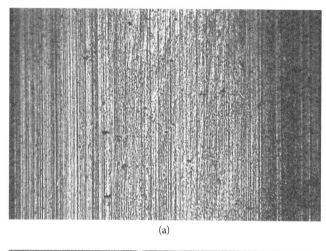

(a)

(b)

FIGURE 5.13
Incipient micropitting at 100 times magnification (a) in a ball bearing race (b) used for a long time in a 16,000 rpm die grinder.

bearing is working properly, the balls do not touch the race. They ride on a separating film of oil—hydrodynamic lubrication is said to exist. However, at start-up and shutdown, there is insufficient speed to generate a hydrodynamic film and metal-to-metal contact can occur. Every metal has a fatigue strength in tension, compression, or shear. If the operating stress is above one of these, failure is likely. Every rolling element bearing has a load capacity that was established by the manufacturer based upon the materials of construction (strength) and the size of the real areas of contact (stress applied). If a bearing is used below its load capacity and within the speed, cleanliness, and lubrication guidelines of the manufacturers, micropitting should not be a problem and the bearing should have a life of 10^{10} cycles, which is usually considered infinite life.

Needless to say, optimum service conditions often do not exist; micropitting can and does occur. For example, if a spindle operates at 20,000 rpm, 10^{10} cycles only takes a few months. Thus, severe operating conditions need to be a factor in avoiding micropitting. Micropitting is common in rolling element bearings, but not desirable. The fractured material gets rolled over by the rolling elements, and this exacerbates the rolling contact fatigue situation.

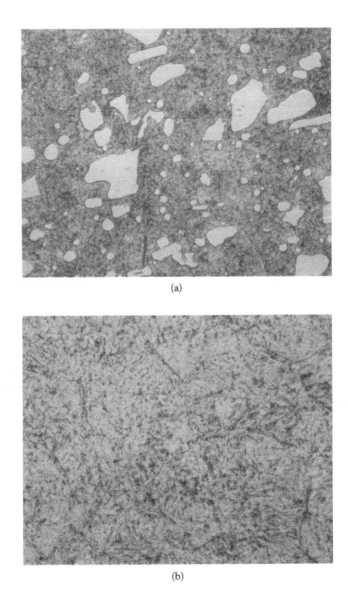

FIGURE 5.14
Microstructure of (a) type D3 and (b) type S7 tool steels. The former contains massive carbides; the latter does not. Both photos are magnified 1000 times.

Micropitting also occurs on gear teeth, usually at the pitch line. The cause is the same as in rolling element bearings—subsurface stresses above the fatigue strength of the gear teeth. Reducing the modulus of one member, like using a cast iron gear versus hard steel, increases the contact area and reduces the stress, but the cast iron has a much lower fatigue strength than a hard steel and is also fraught with inclusions in the form of graphite. Therefore, there are often no "super material" answers to micropitting. Contact stress reduction is the most effective—larger bearings, lower loads, lower speeds, etc. Micropitting can occur on one surface in the rolling tribosystem or on both surfaces. It is never desirable to see micropitting in any rolling tribosystem.

Pitting

Pitting is material removal by the formation of macroscopic craters in surfaces subjected to rolling contact fatigue conditions (Figure 5.3). Pitting is more common in metals that are not hardened to maximum hardness or metals like bronze or cast iron that may not be hardened. Pitting is common on steel rails and wheels used in railroads and trolleys. Some "pits" are indents from riding over debris on tracks (like stones), but some pitting is due to surface fatigue.

Steel mill work rolls often suffer pitting from the stresses encountered while cold rolling steels. Pitting can be due to indentations from foreign bodies or from surface fatigue—particles fractured from the rolling surface. Work rolls used on products at elevated temperatures can have cracks caused by thermal fatigue as well as pits due to compressive stressing. The usual solution to pitting in rolling tribosystems is to increase the strength of the contacting surfaces and reduce the loads.

Most railroad wheel/track systems cannot be lubricated because of traction needs, but hardened surfaces can be employed to reduce pitting problems. Railroad wheels and track surfaces can be surface hardened. It is also common to use steels like Hadfield steels, which increase in strength with repeated cold reduction (working) from rolling. Figure 5.15 shows a work-hardened surface of a 13% manganese Hadfield steel used in a battering application. The surface hardness went from 270 to 460 HV from the surface cold working. These steels are frequently used for railroad frogs (switch gaps) where rolling impact is common. The cold working reduces wear and deformation of the rolling surfaces. It should be kept in mind, however, that these steels do not get harder and stronger without plastic deformation. They must change shape to get stronger. These size changes may alter their ability to function as intended. Hardfacing with fusion welds of special alloys is also used to address pitting in rails. Just about any ferrous metal can be applied by hardfacing, as

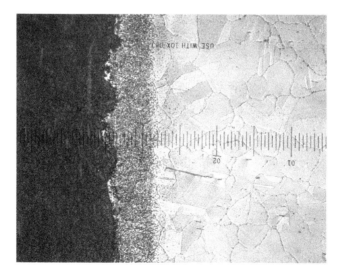

FIGURE 5.15
Cross section (100×) of a cold-worked surface on a 13% manganese Hadfield steel. The surface hardness was 45 HRC; it was 35 HRC 50 μm from the surface, 30 at 100 μm from the surface, and the bulk hardness was 27 HRC.

can special metals like nickel- or cobalt-based metals. Hadfield steels can also be applied as a weld overlay. They reduce pitting by increasing the strength of the material that is subjected to the repeated stresses caused by rolling contact.

Spalling

Spalling is material removal from a solid surface by subsurface fracture initiated by repeated stressing of the surface. Figure 5.1 shows roadway spalling of the type that produces the roadway potholes that populate most roadways in cities in the United States. Repeated rolling of vehicles initiates subsurface cracks; the cracks grow and a piece of roadway spalls. In freeze-thaw climes, freezing water can cause spalls, but some of the worst roadway spalling on our planet exists on the city center streets in San Diego, California. There is no freeze-thaw to blame there, only surface fatigue and possibly inferior (low-strength) concrete. Figure 5.2 shows spalling on a well-used trolley track. Some anomaly created higher than normal stress in that localized area. Or it may be that there were many subsurface cracks in that section of rail and one grew faster than the others (because of statistics) and one spall occurred on an otherwise acceptable section of rail. Figure 15.12 shows spalling of a hard plating on a soft steel shaft subjected to rolling contact from a ball bushing. Coatings are particularly prone to spalling when subjected to rolling contacts. The Hertz stress tends to be highest at the plating-substrate interface, and thus cracking starts there and spalling of the coating is the usual manifestation. It took rolling element bearing manufacturers decades in the United States to develop a plating that would not spall on raceways. Some use what is called thin dense chromium, which is just what the term states. Its thickness is kept to 1 or 2 μm and special treatments (proprietary) are used to enhance the bond to hard steel. Some thin-film coatings are graded in strength and elastic modulus to minimize the ice-on-snow situation that occurs with a hard coating on a soft substrate. Whenever coatings are used in rolling tribosystems, spalling is a major concern to be addressed.

Slip in Rolling Tribosystems

As mentioned previously, rolling only occurs in a fraction of any Hertzian contact. The fact that a ball pressed on a flat surface produces an area not a point of contact means that elastic movement occurred on getting the ball to form an area contact (Figure 5.9). Where there is rubbing there is the potential for sliding wear. So sliding is conjoint with rolling, and rolling tribosystems will always have some sliding wear present. This sliding may show up as polished lines of travel rather than micropitting, or on the other extreme, they can cause severe track and wheel wear. This often occurs on curves in rail/wheel systems. Vehicle tire wear is due to slip at the rolling contact. Ball bearings with metal balls in plastic raceways wear by slip of the balls on the plastic (Figure 5.16). Shopping cart wheels and utility cart wheels usually fail by slip-induced wear rather than surface fatigue. Shoppers move shopping carts in all directions rather than in a straight line forward to progress to the meat section.

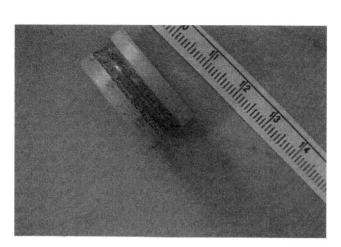

FIGURE 5.16
Adhesive transfer from stainless steel balls to an acetal plastic (POM) raceway.

Testing Materials for Rolling Fatigue Resistance

The traditional test for the rolling contact fatigue resistance of ball bearings is to simply static load a bearing and run it to failure (Figure 5.17). This may require months and many bearings to establish Wiebull statistics for a critical load that the bearing can withstand. Acoustic emission devices on test bearings can be used to detect the onset of damage—like micropitting. Bearing testers can also be heated to simulate hot operating environments or cooled to simulate cold environments. More quantitative lab testing of bearing materials can be done with mating disk specimens. Both can be driven independently to produce various degrees of sliding to rolling contact. Surface micropits or pits are often the test metric.

Wheels for trains are tested on full size wheels on a dynamometer or on trains in test tracks. There are few material options for rails and train wheels, so these tests are not

FIGURE 5.17
Ball bearing test rig.

widely used. Tires are tested for rolling wear, but since all tires are made commercially, testing techniques and wear data are usually confidential. Spalling and surface fatigue are usually not issues with rubber tires, but tread wear is, and most tires are rated by their manufacturer for relative tread wear life within a manufacturer's product mix. Roadway fatigue is a huge problem, but there seems to be no work in this area in the U.S. tribology community.

Summary

Rolling contact fatigue is common in many rolling tribosystems. Its origin is in the subsurface stresses produced by Hertzian contacts. It is manifested in various degrees of fracture from small pits (microscopic) to macroscopic pits to spalling (large chunks). If rolling element bearings are going to fail, they usually fail by some form of rolling contact fatigue. If coatings are used in a rolling tribosystem, rolling contact fatigue must be made a priority concern—they are particularly prone to it.

Related Reading

Alley, E.D., Influence of Microstructure in Rolling Contact Fatigue (Thesis), Atlanta: Georgia Institute of Technology, 2002.

Beswick, J., Ed., *Bearing Steel Technologies: Developments in Bearing Steels and Testing*, Vol. 8, ASTM International: West Conshohocken, PA, 2010.

Blau, P.J., Divikar, R., *Wear Testing of Advanced Materials*, West Conshohocken, PA: ASTM International, 1992.

Campbell, F.C., Ed., *Fatigue and Fracture: Understanding the Basics*, Materials Park, OH: ASM International, 2008.

Davis J.R., Ed., *Gear Materials Properties and Manufacture Materials*, Materials Park, OH: ASM International, 1988.

Dowson, D., Taylor, C.M., Godet, M., *Mechanics of Coatings*, New York: Elsevier, 1990.

Hoo, J.J.C., Ed., *Rolling Contact Fatigue Testing of Bearing Steels*, STP 771, West Conshohocken, PA: ASTM International, 2001.

Hoo, J.J.C., Green, W.C., *Bearing Steels into the 21st Century*, West Conshohocken, PA: ASTM International, 1998.

Tourret, R., Wright, E.P., *Rolling Contact Fatigue Performance of Lubricants*, London: Institute of Petroleum, 1977.

COLOR FIGURE 2.7
Excrescence formation on test specimens in the ASTM G 98 galling test. This damage occurred in one 360° rotation of the upper button on the lower counterface or block. The left test was performed at an apparent contact pressure of 5 ksi (35 MPa) and the right at 3 ksi (21 MPa).

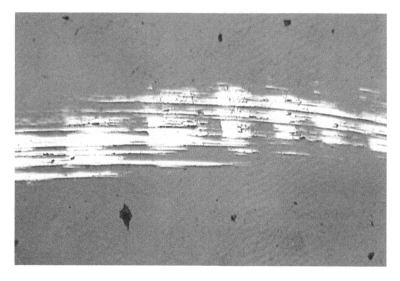

COLOR FIGURE 3.2
Adhesive transfer of titanium to glass by one forcible rub (100×).

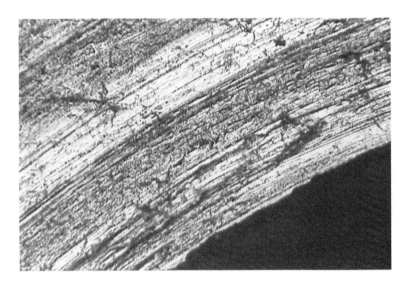

COLOR FIGURE 3.5
Incipient galling (100×)—microscopic excrescences instead of macroscopic ones.

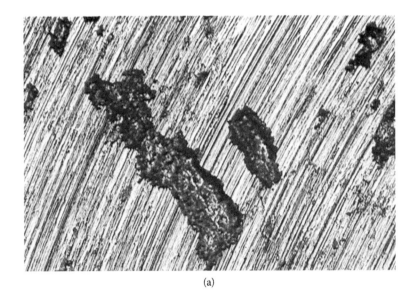

(a)

COLOR FIGURE 3.6
Adhesive transfer of titanium (a) to steel (b) in an ASTM G 98 galling test (100×).

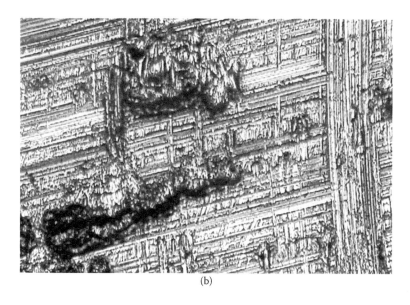
(b)

COLOR FIGURE 3.6 (continued)
Adhesive transfer of titanium (upper (a) photo) to steel (lower (b) photo) in an ASTM G 98 galling test (100×).

COLOR FIGURE 3.9
Balls welded together by adhesive interaction (seizure) in a four-ball wear test for oils. One ball rubs on the lower three, which are held stationary, and the load on the upper ball is incrementally increased until the films separating the balls no longer separate and adhesive interaction (welding) occurs.

COLOR FIGURE 3.13
Polishing wear of hard steel saw components after years of service.

COLOR FIGURE 3.20
Bore of a plastic bushing after a temperature-induced seizure.

COLOR FIGURE 3.21
Adhesive transfer of carbon graphite to a copper commutator from the carbon graphite brushes that rub on it.

COLOR FIGURE 4.1
A stone entry step worn concave by centuries of abrasion from hard particles on footwear.

COLOR FIGURE 4.2
Scratching abrasion from sliding a rock on a finely finished wood surface. (Do not attempt replication of this experiment without permission of the owner of the finely finished surface.)

COLOR FIGURE 4.4
Indentation of a wood surface from rolling contact with a rock.

(b)

COLOR FIGURE 4.14
Scratches at 100× produced by sand (a) and aggregate on the hard steel of a mason's trowel (b).

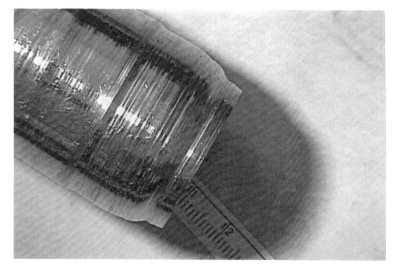

COLOR FIGURE 4.15
Abrasion produced by asbestos packing on a stainless steel bushing.

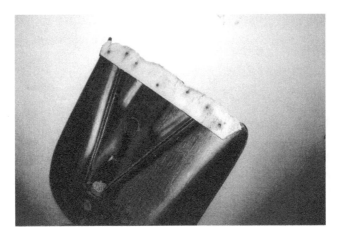

COLOR FIGURE 4.17
Abrasion of a plastic ultrahigh molecular weight polyethylene (UHMWPE) wear strip on a plastic snow shovel.

COLOR FIGURE 4.18
Abrasion of plastic heels on a ski boot from walking in parking lots free of snow.

COLOR FIGURE 5.5
Cracking on a railroad track.

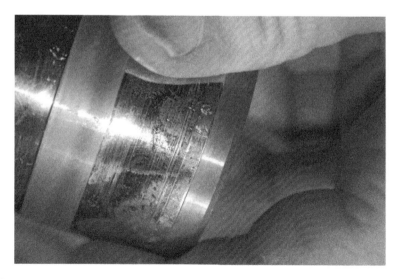

COLOR FIGURE 5.6
Spalling of a hardened steel rolling surface for cam followers.

COLOR FIGURE 5.7
Indentation in a ball bearing raceway from contaminant particles.

COLOR FIGURE 5.8
Brinelling of a raceway from static overload.

COLOR FIGURE 6.3
Impact wear on a cast iron post-driving maul.

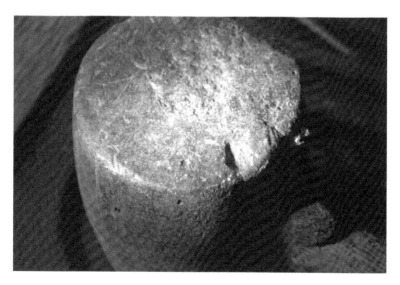

COLOR FIGURE 6.4
Impact wear on the same maul (other end) that was used on steel posts—spalling occurred.

COLOR FIGURE 6.5
Impact wear on a surveyor's sledge after years of driving survey stakes (wood and metal).

COLOR FIGURE 6.10
Impact wear on an elastomer shoe heel.

COLOR FIGURE 6.11
Impact wear on two different rubber hammers.

COLOR FIGURE 7.13
Built-up edge on lathe tool.

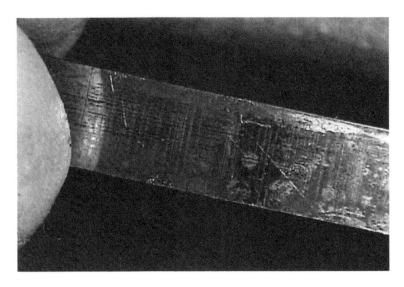

COLOR FIGURE 7.14
Mild wear on a piston ring.

COLOR FIGURE 8.3
Abrasion of flights at the inlet section of a polyester extruder after introduction of carbon black into the melt. The screw was etched to show how the chromium plating had been removed. The flights were 12 mm wide to start, and the eroded width was as low as 2 mm.

COLOR FIGURE 9.5

Effect of impact angle on a brittle material: 20 strikes of a brick hammer at a 20° incidence angle (by thumb) and 20 strikes at a 90° angle to the flat face. Note material removal differences.

COLOR FIGURE 9.11

Typical crater developed in the ASTM G 76 SPE test.

(c)

COLOR FIGURE 10.1
(c) Close-up of droplet erosion from the walkway.

COLOR FIGURE 12.5
Bicycle tire choices. Which has lower friction with hard roadways? Which one in dirt?

6

Impact Wear

Impact wear is progressive material removal or damage to a solid surface caused by repeated compressive stressing of that surface. The wear/damage can affect one or all members involved in the impact action. A hammer used in carpentry wears by impact against nails (Figure 6.1), but there is no concern of damage to the contacting surface, the counterface, or the nails. The nail may receive plastic deformation or even material removal, but nobody cares. The nail head has plenty of material left to do its job of holding member shapes together. However, some applications where impact wear occurs—damage to any of the members involved in the impact—can affect serviceability. A classic example of such a tribosystem is a check valve (Figure 6.2). Balls and other shapes impact a contour seat to produce a fluid seal (air, water, oil, etc.). If the impacted surfaces lose material or form surface protuberances that affect sealing, the device will fail.

Impact wear is also significant in metalworking tools like swaging and cold-heading dies, riveting tools, punch press dies—countless tools that strike metals to form them into shapes. All of the impact tools that a blacksmith uses to shape metal suffer impact wear. Impact wear combined with abrasion is the cause of dulling of tools used to break hard, brittle materials like concrete and rocks. The abrasion component often overwhelms the impact component, but it is still there. The maul shown in Figure 6.3 was designed to drive wood posts. In this application, the impact damage occurred mostly on the fence posts. However, one end was occasionally used to drive steel fence posts, and spalling was the result (Figure 6.4). Figure 6.5 shows the impact damage on a sledgehammer used to exclusively drive metal and wood stakes. In this instance, the entire end fractured due to subsurface fatigue.

One of the least publicized examples of impact wear is on prosthetic devices like a hip joint made from a metal cap and matching metal femur (Figure 6.6). Metal-to-metal human joint replacements are very popular in Europe (200,000/year in Switzerland in 2012) and other regions of the world because they wear much less than metal-on-plastic couples. However, when engaging in physical activities that cause disengagement of the mating members (like some types of exercise or sports activity), the metal surfaces can see impact wear and even spalling. Ceramic-on-ceramic joint prosthetics are also susceptible to impact damage. Needless to say, any spalling or material removal of any kind from these surfaces is undesirable.

Thus, impact wear is present in many tribosystems. This chapter will discuss the mechanisms of damage under impact conditions, damage in plastics, and elastomers, in metalworking and in mineral benefication.

Mechanism

Figure 6.4 shows denting and spalling of material from the impact surfaces can produce damage similar to that produced in rolling fatigue. The mechanism is the same. Subsurface

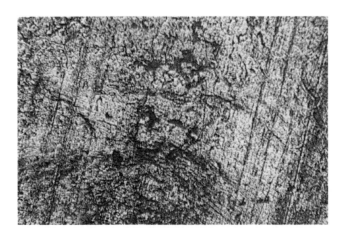

FIGURE 6.1
Typical surface damage on a well-used carpenter's hammer head (100×).

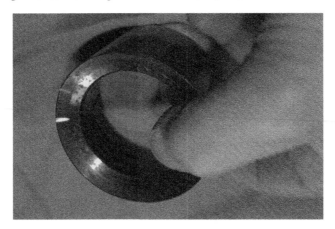

FIGURE 6.2
Impact wear on a valve seat from a hard steel ball contacting twice a minute for thousands of hours.

FIGURE 6.3 (See color insert.)
Impact wear on a cast iron post-driving maul.

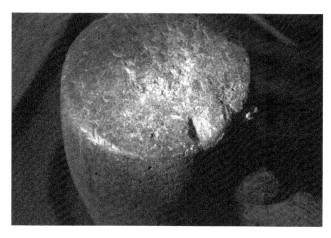

FIGURE 6.4 (See color insert.)
Impact wear on the same maul (other end) that was used on steel posts—spalling occurred.

FIGURE 6.5 (See color insert.)
Impact wear on a surveyor's sledge after years of driving survey stakes (wood and metal).

stress can exceed the fatigue strength of the material; a crack will be initiated, and with continued impact, it will propagate to the surface and produce a spall (Figure 6.7). Impact (like rolling) produces relative sliding on a surface that can remove material by adhesive wear. When one surface impacts another surface, both surfaces see compressive stress as well as relative motion (Figure 6.8). The surfaces of both members may elastically or plastically deform, and material from (a) and (b) will move laterally in radial directions to accommodate the impact stress. Hooke's law dictates that if you apply a stress (force) to a material, it will elastically deform until the stress reaches some level and then plastic deformation occurs.

When a hardened hammerhead strikes a soft nail, the stress on the head is not sufficient to cause plastic deformation of the hammerhead, but is usually sufficient to cause deformation of the impacted surface. Chisels used in auto mechanics often have a soft head, and, by design, it suffers plastic deformation in deference to the hammer (Figure 6.9). The impact ends of cold chisels are tempered to be much softer (and weaker) than the business

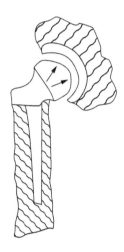

FIGURE 6.6
Schematic of impact wear in a metal-on-metal hip prosthesis.

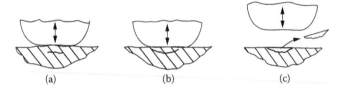

FIGURE 6.7
The mechanism of impact-produced surface spalls.

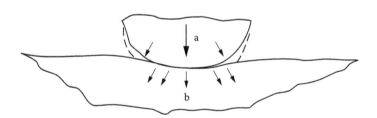

FIGURE 6.8
Surface movement (sliding) under impact conditions. Surface (a) bulges, surface (b) deflects downward.

end of the chisel. This is done because hard–hard impacts can produce spalled material with the velocity of a bullet from a gun. For example, using a mason's hammer on hardened concrete nails can produce a particle that goes through a portion of the hand holding the nail. Of course, eye protection is used, but that may not be enough to stop some spalls that become projectiles. Check valves in hydraulics involving high pressures and impact forces are usually hard–hard couples to minimize plastic deformation, but they usually fail by spalling. However, the spalled projectiles are usually contained and do not create a safety concern, but do cause a contamination concern.

FIGURE 6.9
Impact wear and deformation on a well-used chisel head.

Impact Wear of Plastics/Elastomers

The same mechanism and rules applied to metal-to-metal impacts apply to plastics and elastomers. Compressive stress from impact can result in subsurface cracking, and there is relative slip on surfaces due to lateral motion produced by the elastic properties of the plastic or elastomer. However, there is a difference with elastomers since some of their deformation can be viscoelastic. Viscoelastic behavior occurs when, for example, a material is indented slightly with a spherical shape and one hour later the indent still appears, but 24 hours later, the indent has disappeared. This is viscoelastic behavior—strain recovery is time dependent.

In most impact applications involving plastics or elastomers, this is not a problem or concern, but it could be in, for example, a rubber seal application where after an indent from a contaminant instant recovery is needed, but may not happen depending on the viscoelastic tendencies of the rubber. Some rubbers like conforming foams for pillows and mattresses (memory foams) are designed to exhibit significant viscoelastic behavior (so your sleep dents disappear during the day).

The most costly impact wear of rubber probably is heel wear on shoes. A new pair of rubber heels will become rounded in probably an hour of working on concrete or similar walkway material. The sharp molded edge sees very large stresses when the heel impacts the pavement. The edge is fatigued and rounding continues to progress until the shape change mandates heel (or footwear) replacement (Figure 6.10). Of course, there is a contribution from abrasion, but break-in wear involves significant impact wear. Rubbers wear by fracture and adhesive transfer to contacting surfaces. The cost of wear of footwear heels is estimated to be $10 per person per year or $60 billion/year. Typical impact wear on rubber mallets is shown in Figure 6.11.

High-durometer rubbers like those used in bowling balls wear most by spalling. They are "brittle" by rubber standards and do not have the elastic properties of the "stretchy"

FIGURE 6.10 (See color insert.)
Impact wear on an elastomer shoe heel.

FIGURE 6.11 (See color insert.)
Impact wear on two different rubber hammers.

rubbers. For this reason, they are not usually used (by design) in impact applications. Golf balls are sort of an exception, but their outer layer is usually not considered to be an elastomer. It is usually an olefin or other plastic. Needless to say, these little balls are often damaged by impact from irons, but since the impact is not repetitive (the club never hits in the same spot twice), they do not suffer impact wear, but impact-induced fracture. They get sliced by irons that do not strike the ball the way the game was intended (however, I still blame ball slices on defective ball covering).

Rigid plastics are frequently used for dishes and cooking vessels. In this application, they are impacted repeatedly in stirring, beating, spoon rubbing, whisking, etc. The impact damage ranges from spalling to dulling, scratching by adhesive transfer of polymer to the impacting spoon or kitchen tool. The thermoset plastics commonly used for dishes are particularly prone to impact wear; these are brittle plastics.

Impact Wear in Metalworking

Impact wear is prolific in metalworking and forming. Punch presses are basic tools for mass producing sheet metal shapes. Sometimes, parts are "blanked" from flat stock (Figure 6.12). The punch and die edges that fracture the blank from the sheet, strip, or plate receive tremendous compressive stress when contact with stock occurs. The impact is proportional to some exponent of the punch velocity (at least a power of 2), and this means very high stresses on the edges of the punch and die. Very often these tools are "sharpened" by surface grinding. This operation leaves a burr, and with the first impact with product stock this burr fractures (Figure 6.13). Microfracture of cutting edges is the usual initiation mechanism in punch and die wear. The fractured edge receives many more impacts (sometimes 1000/min), and microfracture continues to round the die edges.

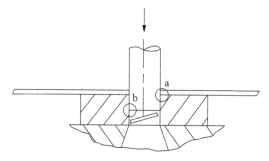

FIGURE 6.12
Sources of impact wear in metal blanking: (a) on the die, (b) on the punch.

(a) (b)

FIGURE 6.13
Impact wear on a tool steel punch: 30× in (a), magnified 1000 times in (b).

This mechanism prevails on all metalworking tools that rely on a cutting edge. There is also lots of adhesive wear from rubbing on stock. Sometimes, there is abrasive wear also. For example, dirt on oily metal can abrade punch and die edges, and sometimes the stock being blanked is abrasive, like plastics coated with abrasive inks. Magnetic particles in or on plastics make the abrasive to many metals.

Hardened tool steel, cemented carbides, and cermets are typical tool materials. These materials have the highest compressive strength of commonly available tool materials, but they all are subject to microfracture of edges, mostly because they contain hard second phases in a weaker matrix. They are heterogeneous in microstructure, and these hard phases act as stress risers to exacerbate the stresses in the material from impact. Generating a cutting edge without any burrs (atomically sharp) will prevent burr fracture, but mostly we do not know how to do this. Riveting tools, forming dies, impact extrusion dies, and swaging dies impact mostly soft metals, and if they do not have a cutting edge, they mostly wear by the rubbing or adhesive component. Just like the example of a hammer hitting a nail, the rubbing comes from the sideways sliding of the nail (by deformation) on the hammerhead. A hard metal sliding on a soft metal in this type of tribosystem usually results in a low wear rate on the hardened tool.

More often than not, the soft metal transfers to the hard tool surface and protects it from sliding wear. If a hard steel hammer is used to drive brass tacks, the hammerhead will take on a brass hue from adhesive transfer.

Thus, impact wear is a significant concern in metalworking tools, and it becomes a limiting factor on tools that need to maintain sharp edges. Impact-induced microfracture of edges is the culprit. The adhesive component of the impact wear can usually be tolerated, but sometimes not, like in impact extrusion of beverage cans. The adhesive transfer to the punch sidewalls often shuts down the process.

Impact Wear in Mineral Benefication

Only two metals come out of the earth's crust in metallic form: gold and copper. The scarcity of gold is well known, and mass copper is essentially extinct. Most copper is obtained from copper ores that may only be 5% by mass copper. Ninety-five percent is "rock" that has to be dealt with. Gold is often detected in sandy deposits, and they often can simply sieve and slosh the soil with water to extract metallic gold. Mass copper was entwined in rocks, and the rocks had to be broken to get the mass copper. Forging-type presses were used to break the rocks and extract (by hand) the mass copper. Breaking rocks in refuse produced impact wear on the punch of the shredder shown in Figure 6.14. The stamp presses used in copper mines used a stamping head attached to a board that raised the head and at a height moved away to allow the heavy head to crush the rock. The rock was fractured to particle form that was called stamp sand.

This same process is still performed in many more sophisticated machines, but all involve some type of impact with hard brittle substances—rocks or minerals of some form. A common rock crusher is a hammer mill that has a central shaft with the hammer attached that strikes rocks dumped into a hopper with the rotating hammers at the bottom. Sometimes rocks are loaded into a horizontal cylinder full of steel balls that may be 10 to 15 cm in diameter. After charging the large cylinder, it is sealed and rotated as

FIGURE 6.14
Impact wear on impact edge of a pug mill used to grind industrial waste.

balls drop onto the rocks and crush them. Another variation uses large steel rods to do the crushing. Impact wear occurs on the balls and rods. They get abraded as well from rubbing on the rocks. However, the spalling produced by the impact generates subsurface fatigue, which is usually the root cause for replacement. There are as many rock-crushing devices as there are minerals to be extracted, but most involve some type of impact from the minerals on a steel or cemented carbide protected surface. Materials with high toughness are used as tool materials (manganese steels, shock-resisting tool steels, high binder cemented carbides, etc.).

Summary

Impact wear is very much like rolling fatigue wear in that subsurface cracking from compressive stressing of the surface produces subsurface cracks that propagate and produce pits or spalls that remove material (Figure 6.2). Impact wear is conjoint with adhesive wear when the impact is with a metal, and abrasive wear predominates as the conjoint wear in impact from rocks and other inorganic substances. Impact wear is accepted as the cost of doing business in some applications. For example, jackhammer drill bits are resharpened as they wear and then discarded when they get too short to use. On the other extreme, a $200,000 blanking die can be resharpened a few times, but not many, so there is great interest in uncovering materials that resist their type of impact wear better. Impact wear on these tools usually takes on the form of microfracture of cutting edges (Figure 6.13). Finally, impact wear on battering tools like hammerheads takes on a gnarled/fretting appearance, as in Figure 6.5, if the battering tool is hard. Battering with a soft tool results in mushrooming (Figure 6.9) caused by plastic deformation. All manifestations of impact wear involve repetitive compressive stressing of surfaces coupled with relative motion between impacting members by gross slip or by motions produced by elastic or plastic strains in the members.

Related Reading

ASTM C 944, *Standard Test Method for Abrasion Resistance of Concrete and Mortar Surfaces by the Rotating Cutter Method*, West Conshohocken, PA: ASTM International.

ASTM D 4508, *Standard Test Method for Chip Impact Strength of Plastics*, West Conshohocken, PA: ASTM International.

ASTM G 73, *Standard Test Method for Liquid Impingement Erosion Using Rotating Apparatus*, West Conshohocken, PA: ASTM International.

ASTM G 76, *Standard Test Method for Solid Particle Impingement Using Gas Jets*, West Conshohocken, PA: ASTM International.

ASTM G 81, *Standard Test Method for Jaw Crusher Gouging Abrasion*, West Conshohocken, PA: ASTM International.

Bayer, R.G., *Engineering Design for Wear*, 2nd ed., Boca Raton, FL: CRC Press, 2010.

Engel, P.A., *Impact Wear of Materials*, Amsterdam: Elsevier, 1978.

7

Lubricated Wear

Anything that separates rubbing surfaces can be a lubricant. However, the ASTM lubrication committee's definition is "a substance that separates sliding surfaces and reduces the wear and friction between those surfaces."

The operative words are *reduce wear and friction*. Surfaces can be separated by sand particles and the friction may be lower, but the wear probably will not. Classical lubricants fall into the following categories (Figure 7.1):

1. Liquids (oils and other fluids)
2. Greases
3. Solid lubricants
4. Gases/vapors

Liquid lubricants are predominantly oils, but water, even molten metals (indium, etc.) are used to lubricate some bearings. Greases are sponge-like gels full of interstices that hold oil to be released when needed by rubbing conditions. They are formulated to keep the oil where it needs to be. The sponge material (mostly lithium soaps) is not intended to lubricate—only to hold an oil that will lubricate, and the nature of the oil can be varied. When a grease is used in a rolling element bearing, it is packed around the balls or rollers to some volume fraction (not full). When rotation starts, the friction between rubbing surfaces produces heat and the oil starts to come out of the "sponge" and lubricate the rolling surfaces by fluid film separation.

Solid lubricants are, as the name implies, solid substances that separate rubbing surfaces and reduce friction and wear. This concept was stolen from nature. Waxes are "solids" obtained from many different plant species. Carnauba wax is scrapped from the leaves of a tropical plant. It can be agglomerated in a ball for shipment. Users can make it easier to spread on surfaces with a solvent, but it has very low solubility in most solvents. When wax is applied to a surface, it forms microscopic flat platelets that separate solids when applied to one or both members. Wax lubrication is widely used on food products on surfaces handling foods because they can be edible. Industrial solid lubricants include substances, such as graphite, molybdenum, or tungsten disulfide. These materials have a hexagonal structure that has an affinity to form platelets on a surface like waxes. These platelets may have heights in the micrometer range, and they separate surfaces to prevent wear, and when these platelets are on both rubbing surfaces, they slide on each other. They can do this usually with a friction reduction compared to that of the unlubricated couple.

Gases and vapors can separate and lower friction. In fact, air bearings as a class of bearings operate with friction coefficients (at design speed) that are almost always lower than that of rolling element bearings (<0.1). Sometimes, just vapors from usually organic substances are used. For centuries, clockmakers have lubricated the brass, bronze, and steel sliding parts in the clock mechanisms by keeping a small open container of kerosene inside the clock enclosure. The kerosene vaporizes and condenses on all surfaces in the

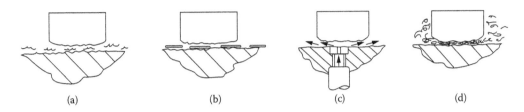

FIGURE 7.1
Types of lubrication: (a) = liquid, (b) = solid film, (c) = pressurized fluid, (d) = gaseous film.

enclosure. Thus, gears, pawls, escapements, and the like see a continuous film of a sub-stance that tends to separate the surfaces to prevent adhesive wear and also lowers friction if the right evaporant is used. This type of lubrication (vapor phase) is still used on mecha-nisms that are difficult to lubricate with oils and greases (because of their mess).

A key concept with regard to lubricated wear is the degree of surface separation pro-duced by a lubricant. The most famous graph in lubrication, the Stribeck curve, explains the separation and friction situation (Figure 7.2). At low sliding speeds, oil-lubricated sur-faces partially or intermittently touch, and the friction is high because the grease does not effectively separate the rubbing surfaces. Thus, low speed produces what is called boundary lubrication—not quite lubricated. As speed increases, the separating film gets thicker and surface separation is more often. This is called the mixed lubrication area. As speed and oil viscosity increase, the oil can produce constant separation of the rubbing surfaces and the friction is more constant and low. This is called hydrodynamic or fluid film lubrication.

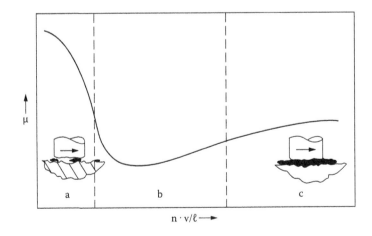

FIGURE 7.2
The Stribeck curve, which shows the effect of a separating lubricant film on the system coefficient of friction. In region (a) boundary or partial lubrication exists; in region (c), the sliding members are separated by a full oil film; in between (b) there is mixed lubrication, sometimes partial film, sometimes full film. n = dynamic viscos-ity of the oil, v = sliding velocity, ℓ = load, μ = coefficient of friction.

In all lubricated systems, if real and constant separation of surfaces is achieved, the rubbing surfaces will not wear; they do not touch each other. However, this utopian situation is not always achievable. For example, all bearing surfaces rub until separation speed is achieved, and they rub again at shutdown. Reciprocating systems go to zero velocity at the end of every stroke, and surface contact may occur. Air bearings can stop with disastrous results if an event occurs that overpowers the supporting air film, like a shock load. With this concept in mind, we will describe what wear looks like on some of the more common applications where lubricated wear prevails:

- Reciprocating internal combustion (IC) engines
- Plain bearings
- Rolling element bearings
- Metalworking
- Machining of metals

Reciprocating Lubricated Wear

Internal combustion engine wear may be a family's highest annual cost next to shelter. In the United States, the average cost of a new automobile is $30,000. Even if it lasted 10 years, the annual cost per vehicle is about $3,000, and vehicles are usually discarded because of lubricated wear:

- Engine ring/cylinder wear
- AC compressor wear
- Water pump wear
- Valve stem wear
- Transmission wear

Of course, tire wear and brake wear are added costs. Cylinder/ring wear is the most costly repair of this group. It means a complete rebuild or engine replacement. This is so costly in the United States that it is seldom done. Cylinder/ring wear causes loss of compression and burning of oil—neither is desirable. Lubricated couples, even under ideal conditions, wear during the boundary lubrication portion of operation. The rubbing members contact. A well-designed reciprocating tribosystem operating under ideal conditions will wear by just changes in surface texture of the rubbing surfaces. In the case of internal combustion (IC) engines, the cylinders are usually cast iron and the piston rings are chromium or thermal spray-coated, hard, white iron. The coatings on the rings smoothen and eventually wear away. The cylinder wears by smoothing of the honed surface, and eventually the wear profile replicates the speed/acceleration profile of the piston. If some event or chronic condition impedes the flow of lubricant to the rubbing surfaces, "scuffing" can occur. We prefer to call it scoring, but significant damage can occur to the rubbing surface in minutes in normal driving (1500 strokes/min). Scoring (scuffing) is plastic deformation of rubbing

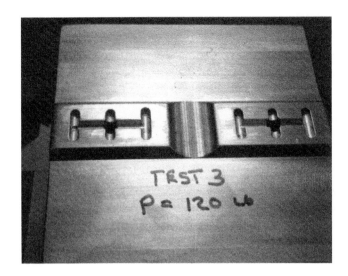

FIGURE 7.3
Flat pad babbitt bearing after reciprocating test with full film lubricant separation.

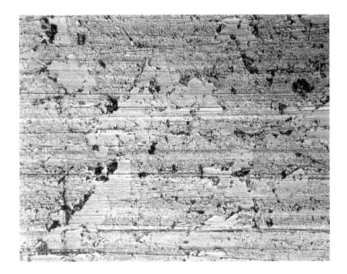

FIGURE 7.4
Burnished babbitt surface after testing under hydrodynamic lubrication conditions (400×).

surfaces from adhesive interaction between the two surfaces. So under ideal conditions, a hard-soft couple wears mostly by lowering the surface roughness of the rubbed area.

Figure 7.3 is a flat pad lead-tin babbitt bearing where lubricant was pumped into the groove pattern. The counterface was hardened steel (60 HRC) with a surface roughness of 0.1 to 0.2 μm. After 10,000 rubbing cycles with sufficient fluid pressure to ensure hydrodynamic lubrication, the rubbed babbitt showed no wear damage (Figure 7.4). When the lubricant pressure was reduced (same load and stroke), scoring (or scuffing) occurred (Figure 7.5). Most soft bearing materials will do this.

FIGURE 7.5
Scoring of babbitt in Figure 7.3 when lubricant pressure was reduced to produce boundary lubrication (50×).

FIGURE 7.6
Tribofilm of plastically deformed metal on a brass bearing that saw counterface contact (400×).

If both rider and counterface are the same hardness (high hardness), the rubbing surfaces will usually become polished in the rubbing areas. The polishing will match the travel pattern. The wear is simply lowering of the surface texture parameters (roughness, etc.). When hard–hard couples lose lubrication and slide on each other, oxidative wear occurs. The steel surfaces will look like they rusted. When a hard sliding member slides on a soft counterface and loses lubrication separation, besides the scoring that can occur on the soft surface, the soft metal surface can be plastically deformed to produce a tribofilm that can be a mixture of both metals or a highly deformed layer of the soft metal (Figure 7.6).

Plain Bearings

A plain bearing is simply a cylinder or other shape that supports and guides a moving shaft or other shape. The most common plain bearings are cylinders used to support a rotating or reciprocating shaft (Figure 7.7). The plain bearing system is used in rotating, oscillating, or reciprocating modes, and the wear pattern on the rubbing surfaces will reflect the loss of fluid film support at the ends of reciprocating motions. Plain bearings can carry heavier loads than rolling element bearings, and they do not require the speeds that rolling element bearings require to get lubricant film separation of surfaces (hydrodynamic lubrication). For example, plain bearings (oil-impregnated powdered metal (P/M) bronze) are commonly used on fractional horsepower (HP) electric motors that in the United States run at about 1700 rpm. Fractional HP motors that run at the next commonly available speed level, 3600 rpm, usually are sold with rolling element bearings. The lower speed limit may be inadequate to achieve balls or rollers separated from the raceways by a film of oil. Overall, in lubricated wear, the goal should be to get a separating film to keep the mating surfaces from contacting. If they do not contact, they will not wear each other.

Many times a reciprocating member is best made in a shape other than cylindrical, like the flat pad in Figure 7.3. It is common practice to call these flat pad or way bearings. Also, plain bearings do not have to be round. They can be any shape, but cylindrical ones are the most used. A synonym for cylindrical plain bearings is *bushings*.

What does wear look like on these bearings? Figure 7.8 shows normal wear on a well-used P/M bronze bearing from a hand tool. There is a polishing/burnishing of the surface, and the pores in the powdered metal tend to close up. If the oil supply in the P/M bearing is too low, scoring will result. This same situation occurs on flat pad bearings. It is common to make large flat pad and plain bearings from a soft metal like tin-lead babbitt. This material is intentionally very soft, so if dirt particles get into the rubbing contact, they will embed in the soft babbitt rather than produce scoring. However, if these soft metal bearings are depleted of oil, they will readily score or scuff (Figure 7.5). Anything that uses up the required clearance will cause failure by seizure (like the excrescence in Figure 3.4 that seized the splined shaft). The required clearance (Figure 7.9) is a function of the mating materials, the lubricant, and the operating conditions. Temperature effects must also be

FIGURE 7.7
Typical plain bearings for rotating and sliding shaft applications.

FIGURE 7.8
Wear on a well-used powder metal oil-impregnated film bearing: (a) before use, (b) after significant use (100×).

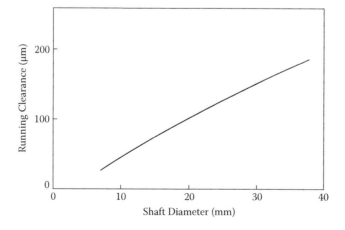

FIGURE 7.9
Running clearance required between a shaft and a plastic plain bearing (overall).

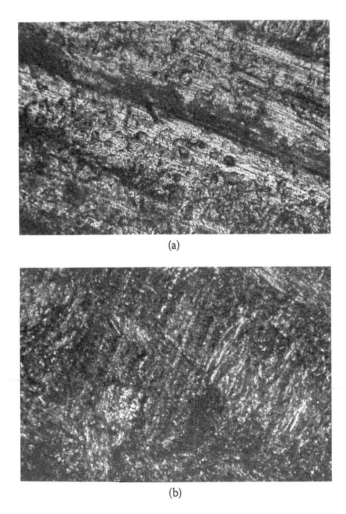

(a)

(b)

FIGURE 7.10
Surface texture changes to a phenolic surface (a); after reciprocating wear (b) (100×).

considered. This is especially important with plastic plain bearings. If they get hot, they expand at least 10 times the rate of the surrounding metal and the net effect is the loss of running clearance. Temperature expansion caused the plastic bush–steel shaft failure shown in Figure 3.20. Plastics tend to produce larger wear particles than metals, and thus surface texture changes after wear tend to be greater (Figure 7.10).

Rolling Element Bearings

Normal wear on well-used ball rubbing surfaces looks like Figure 7.11 The only manifestation of wear is lowering the surface roughness in the rolling path. The balls are polished as manufactured so their surface roughness changes are not perceptible by ordinary surface texture measurement techniques. If a bearing ran dry (no lubricant) for some length

FIGURE 7.11
Lines of travel on a hard shaft running against hard balls in ample lubrication.

of time, the surface texture change would be much greater and micropitting or spalling could occur.

Corrosion is a common cause for failure of rolling element bearings. There are many industrial applications that involve the use of rolling element bearings near water or other corrodents. If a corrodent inadvertently finds its way into the oil in a rolling element bearing, it will be emulsified while the bearing is running, but during downtimes, the corrodent will pit the balls/rollers and the raceway, and when operation resumes, the corrosion product (rust) will act as a solid particle contamination and promote micropitting in addition to the corrosion pits, and failure is eminent. Boat trailers are a prime example of "water in the bearing." The trailer wheels can carry significant loads, and when used, they are submerged in water to off-load the boat. Some have seals that work, some do not. Water gets into the bearing, and it sits there for a week or two until the next use. Then at 70 mph, a noise may emanate from the following trailer. It may be the rolling element bearing converted to broken pieces from surface fatigue/spalling (as happened two times to this author). Needless to say, rolling element bearings must be kept free of materials that might contain solid particles and substances that might cause corrosion of the rolling elements or raceways. Even condensation can cause enough moisture to make an oil corrosive. A simple test for corrosivity of oil in a sump is to simply leave a clean piece of finely ground steel half submerged in the sump for 24 hours. If rusting occurs in the submerged part of the test strip (or in the out portion), there can be a problem waiting to happen.

Metal Forming

We already mentioned the use of waxes to minimize wear on contacting surfaces, but lubricants are needed in many metal-forming processes to minimize adhesive transfer of the metal to the tools. For example, when a cup is drawn from a sheet metal blank as shown in Figure 7.12, the sheet metal rubs with great force on the walls and sides of the

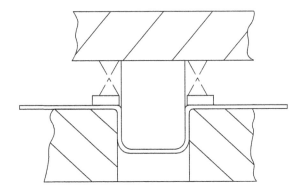

FIGURE 7.12
Drawing a metal cup from sheet metal with a punch and die.

punch. If the metal adhesively transfers to either surface, it will tend to build up and use up the punch-to-die clearance, and more importantly, the cup may show marks or even fracture during the draw. To prevent these undesirable occurrences, the sheet metal blank is lubricated. There are special lubricants for each type of forming operation, but when they do not work, the result is adhesive transfer or galling.

Adhesive transfer to tools is also a concern in rolling; most low-carbon steel sheets for making cans and the like cannot even be manufactured unless their front and back surfaces are coated with oil. They are not dripping wet, but the oil is there (a teaspoon of dioctyl sebacate lubricant per football field of steel in one tin mill operation). These rolling lubricants minimize adhesive transfer to rolls. These oil films are so thin that they usually do not require removal for painting.

Aluminum foil, the kind we wrap our food in, is often lubricated with rapeseed oil on the shiny side, and this is an edible lubricant. Most of the sheet metal parts in auto bodies need an application of lubricant to keep them from adhering to tool surfaces. Thus, adhesive transfer is the issue in metal forming. Lubrication is needed to minimize it.

Machining

The adhesive transfer to tools is also a significant problem in cutting tools (Figure 7.13). Lathe bits, milling cutters, drills, taps, etc., are subject to adhesive transfer from the work. For example, it is almost impossible to tap a small tapped hole in aluminum without a tapping fluid. The soft aluminum will adhesively transfer to the tap, use up the clearance, and most often result in tap breakage. Edge buildup like that illustrated in Figure 7.13 causes the cut to be disrupted and the metal removal drastically reduced. Buildup on drills leads to breakage.

The heat generated in machining operations adds to the adhesive wear problem. When the work gets hot from rubbing friction, it gets softer and adheres to the tool easier. In lathe operations, it is common for cutting chips to come off the work red hot. Machining lubricants are usually designed to be emulsified in water. The water cools the work to reduce its tendency for the material being cut to adhere to the tool, and the oil or other lubricant in the cutting fluid assists by trying to form a film to separate chips from the tool. Adhesion will not occur if the chips are not allowed to touch the tool. This adhesive transfer is

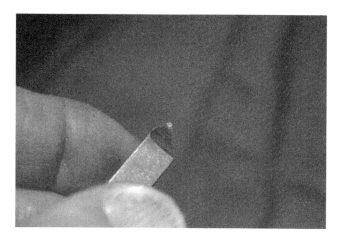

FIGURE 7.13 (See color insert.)
Built-up edge on lathe tool.

also a problem in machining plastics. Fortunately, plastics are usually molded rather than machined to shape. Thus, the problem is not as severe, and machining lubricants for plastics are not as available as metal machining fluids.

Summary

Lubricated wear in properly designed bearings running under ideal conditions results in a mild lowering of surface texture features in the travel path (Figure 7.14). When something interrupts the lubricant scuffing/scoring can result (Figure 7.15) on plain bearings and micropitting/pitting/spalling in rolling element bearings.

In metalworking and machining, wear takes on the form of adhesive transfer to the tools (Figure 7.13). This needs to be minimized to keep the workpiece within specifications.

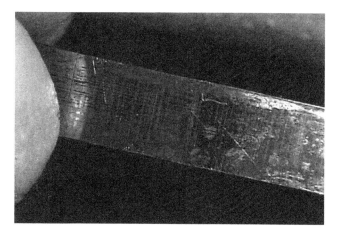

FIGURE 7.14 (See color insert.)
Mild wear on a piston ring.

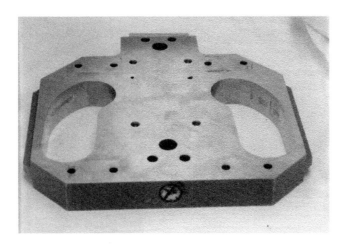

FIGURE 7.15
Scuffing in cam paths.

Related Reading

ASTM D 4170, *Standard Test Method for Fretting Wear Protection by Lubricating Greases*, West Conshohocken, PA: ASTM International.

ASTM G 133, *Standard Test Method for Linearly Reciprocating Ball-on-Flat Sliding*, West Conshohocken, PA: ASTM International.

Berthe, D., Dawson, D., Godet, M., Taylor, C.M., *Fluid Film Lubrication—Osborn Reynolds Centenary*, Amsterdam: Elsevier, 1986.

Booser, E.R., Ed., *CRC Handbook of Lubrication: Applications and Maintenance*, Vol. 2, Boca Raton, FL: CRC Press, 1983.

Booser, E.R., Ed., *CRC Handbook of Lubrication: Theory and Design*, Vol. 1, Boca Raton, FL: CRC Press, 1984.

Booser, E.R., *Tribology Data Handbook: An Excellent Friction and Wear Resource*, Boca Raton, FL: CRC Press, 1997.

Ludema, K.C., *Friction, Wear, and Lubrication: A Textbook in Tribology*, Boca Raton, FL: CRC Press, 1996.

Rigney, D.A., Ed., *Fundamentals of Friction and Wear of Materials*, Materials Park, OH: ASM International, 1987.

Sethuramish, A., *Lubricated Wear*, Amsterdam: Elsevier, 2003.

8

Tribocorrosion

Tribocorrosion is not a very old word. It started to appear in the literature in the 1990s. Its definition is "degradation or transformation of solid material due to the combined effect of corrosion and wear."

The corrosion can exacerbate due to the wear (rubbing) or be less than in a nonrubbing situation. If a plus or minus effect is due to the rubbing, synergy is said to exist. There is an ASTM test (G 119) quantifying the synergy in a tribocorrosion system. There are lots of details involved, but the way this works is: run a slurry abrasion test in a given slurry and measure the mass loss and volume loss; repeat the test with the slurry buffered so that it is noncorrosive to the test specimen. The difference between these results is a measure of the synergy. In most cases, the rubbing action on a surface increases its corrosion rate.

Because corrosion in metals is electrochemical in nature, the chemical reactions during wear can be studied in detail with specialized corrosion measuring equipment, most notably a potentiostat. This is the classic tool used by corrosion engineers to deduce the nature of the corrosion processes. Its use is rather complicated, so we will describe it in more detail, and then discuss important tribocorrosion problems like slurry erosion, prosthetic devices, and common sliding systems (metals and plastics) that operate submerged in liquids that have the potential to cause corrosion to the couple involved in the rubbing.

Use of Potentiostats to Study Tribocorrosion

The slurry pot test that has been used in developing the ASTM G 119 wear/corrosion synergy standard uses mass-only change to explore the wear/corrosion synergy. When electrochemical corrosion monitoring techniques are used, they can provide additional insights into a tribocorrosion process. For example, most corrosion-resistant metals derive their corrosion resistance from a passive surface film that usually forms spontaneously over some time period. Potentiostats can provide information on the time required to reform after a rub. They can establish the kinetics of the corrosion component of tribocorrosion. It is even possible to determine if a voltage (and how much) can be applied to the rubbing couple to stop the corrosion wear synergy.

The basis for the potentiostatic or potentiodynamic corrosion studies is the fact that metallic corrosion involves a potential between mating materials immersed in a electrolyte, and when corrosion is taking place, an electrical current will flow between the members that is proportional to the rate of the corrosion. The current (flow of electrons) comes from oxidation and reduction reactions at the anode—the corroded member and the cathode, the protected member. The oxidation-reduction reactions are below:

Corrosion → oxidation at the anode reduction at the cathode:

Oxidation: $M \rightarrow M^{n+} + ne^-$

$M + n\,H_2O \rightarrow MO_{n/2} + H + ne^-$

Reduction: $2H^+\,2e \rightarrow H_2$

$O_2 + 2H_2O + 20$

where:
n = metal atom
e = electron

A laboratory set up to study the tribocorrosion of a rubbing couple might look like Figure 8.1. The potentiostat is a piece of electronic hardware that monitors the voltage (potential) and current (electron flow) in a corrosion reaction. The higher the potential and current, the higher the corrosion.

The rubbing pair is submerged in whatever solution is involved in the tribosystem, and the system potential and corrosion current is monitored before and after rubbing. A polarization curve produced by the instrument may resemble Figure 8.2. The corrosion potential increases when rubbing starts and goes back to normal when rubbing stops. Corrosion engineers have used polarization techniques to arrive at electronic systems for stopping electrochemical corrosion of submerged boat parts, and it has been demonstrated that this concept can be used to stop tribocorrosion synergy and possibly most of the wear process—the part that involves an oxidation reaction like freshly cleaned iron with air.

In any case, potentiostats are the tool used by tribocorrosion researchers to find ways of dealing with exacerbated corrosion when surfaces rub in a corrosive environment.

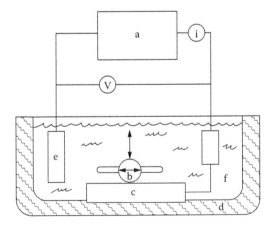

FIGURE 8.1
Schematic of a tribocorrosion test rig using a potentiostat (a), (e) = reference electrode, (b) and (c) = rubbing test couple, (f) = solution of interest.

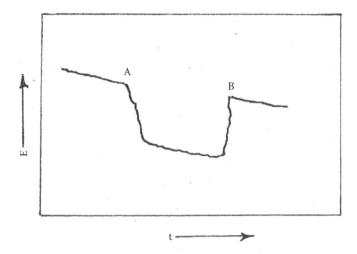

FIGURE 8.2
Typical polarization curve from a tribocorrosion test rig using a potentiostat: (a) = start of rubbing, (b) = end of rubbing, (E) = corrosion potential (voltage).

Slurry Erosion

A slurry is a pumpable mixture of solids and liquids. "Wet" concrete can be pumped, so it can be considered a slurry, even though this "slurry" is 90% or more solids. On the other end of the scale, clear water out of a pristine forest stream may have some suspended minerals in it, but it would not be considered a slurry. From the practical standpoint, a slurry that can cause erosion in an industrial situation should have at least a fraction of a percent of solids by weight (not parts per million), and the solids are intentional or known to be there. In addition, the particles have to have a size that is sufficient to produce erosion. A 10% by weight of 1 µm aluminum oxide in water is a common metallographic polishing slurry. Solids like glass and carbon particles in molten plastic can be considered to be slurries if they have at least a fraction of a percent and enough size to produce damage in rubbing on solid surfaces. Silica particles down to 5 nm in size have demonstrated abrasivity in oils, so there may not be a lower size limit below which abrasion will not occur.

The preceding discussion is evidence that slurries are ill defined. Maybe a realistic definition is a pumpable mixture of solids in a liquid without size limits and percentages. Plastics can still be included since they behave as liquids at the temperature range where they are pumpable. They certainly can cause erosion, as shown in Figure 8.3. This extruder screw had been operating in a specific polymer for almost 10 years with negligible wear. When a fraction of a percent of submicron carbon black was added to the original polymer, severe wear occurred on the flights in the feed section of the screw within months after

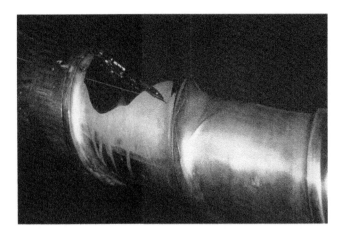

FIGURE 8.3 (See color insert.)
Abrasion of flights at the inlet section of a polyester extruder after introduction of carbon black into the melt. The screw was etched to show how the chromium plating had been removed. The flights were 12 mm wide to start, and the eroded width was as low as 2 mm.

adding the carbon black. The nature of the material removal was progressive polishing away of hardened steel.

The erosivity of concrete, the "fully loaded" slurry, is very low on chutes and mining equipment—mostly scratching abrasion that can be tolerated (Figure 8.4). Thus, another factor that controls the ability of a slurry to remove material (cause wear) is the force with which the material contacts rub the surface. Velocity is also part of the force since momentum forces usually have the form of mv^2, mass times velocity squared. Finally, slurries can have a corrosion component. This is even true for molten plastics. For example, PVC plastics are known to generate HCl vapors, which can be extremely corrosive.

Slurries become an erosion concern when they are handled—conveyed, pumped, directed, sprayed, etc.—any time that they are in contact with a solid surface with significant velocity. Particle jet-cutting rigs using 10% garnet in water can cut through 15 cm of solid glass at a rate of centimeters per minute. The pressure is usually about 50,000 psi (344 MPa) in these cutting rigs, and the particles may have a strike velocity of 100 m/s. The corrosion component is nil because of the speed of the operation. Coal and other mineral slurries that must be pumped from mines to processing facilities or even simply transported to market in slurry form are important slurries from the economic standpoint. Sometimes, coal is transported hundreds of miles (from mine to user) in slurry form, and there are concerns about the slurry eroding through the pipe and ruining the slurry pumps. The usual appearance of slurry erosion is often polishing. If the particles are fine enough, the material will be removed by polishing rather than scratching—like the extruder screw in Figure 8.3.

Sometimes, the corrosion component of a tribocorrosion system is stronger than the abrasion component. Figure 8.5 shows chemically assisted erosion of an anodized coating on an aluminum gate. People's hands carry dirt that can abrade, but they also carry body fluids that can be corrosive (for example, fingerprints usually contain uric acid and sweat contains salt). Probably the corrosion component of hand rubbing contributed more to coating removal than the abrasive substances on people's hands.

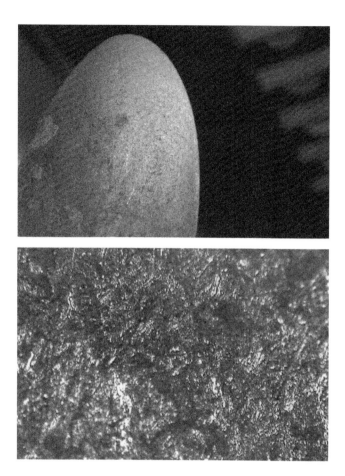

FIGURE 8.4
Slurry erosion on a concrete finishing tool. Damage is predominantly scratching abrasion (100×).

FIGURE 8.5
Tribocorrosion of an entry gate (anodized coating on aluminum) from rubbing produced by user's hands.

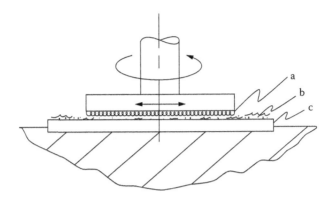

FIGURE 8.6
Schematic of chemical–mechanical polishing (CMP): (a) = CMP pad, (b) = slurry, (c) = silicon wafer.

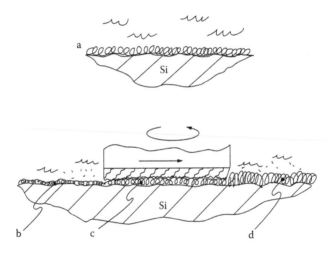

FIGURE 8.7
Mechanism of CMP: (a) and (d) = chemical reaction film, (c) = polishing of film, (b) = film reforms after polishing.

One of the most important industrial applications of slurry erosion is chemical–mechanical polishing of silicon wafers and integrated circuits for computer chips and other electronic hardware. A slurry is made with chemicals that attack the part surface. The solids in the slurry can be abrasives like silicon carbide and aluminum oxide. A rotating polymer pad imposes the slurry on the wafer or part surface and it is planarized—made flat and smooth by the action of the chemical and the slurry (Figure 8.6).

Material removal can be quite fast (several micrometers per minute). The reason this process is called chemical–mechanical planarizing (CMP) is that use of chemicals to produce a reaction film is key to the process. The abrasive would eventually polish the silicon, but the process is greatly speeded by putting chemicals in the liquid phase that cause soft-reaction product films (like oxides); then the abrasive easily abrades the films and the process repeats (Figure 8.7). CMP uses the synergy effect of a chemically reactive slurry medium to speed the erosion. This process employs tribocorrosion.

Another example of intentional slurry erosion is the slurry extrusion process. Abrasive particles are mixed into a polymer gel-like material (like putty). This gel-like material is

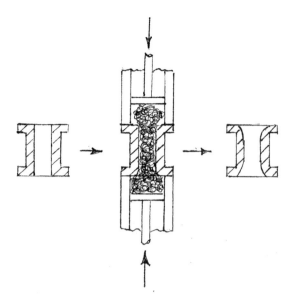

FIGURE 8.8
Using slurry erosion to smooth and polish inside diameters.

forced back and forth through parts to polish, deburr, or sometimes shape (Figure 8.8). A charge of media is hydraulically forced through the hole in the part. When most of it is through, the process is reversed and the abrasive slurry is pushed back down. After a designated number of reversals, the bore of the part will be polished and the sharp edges on the inside will be radiused. Many times in industry, internal passages must be polished and radiused, and this special machine allows engineered slurry erosion to do the job.

Application of "rubbing compounds," liquid polishes for metals, and even the use of toothpaste are examples of intentional slurry erosion. Rubbing compounds are abrasives in a wax-like medium. Liquid metal polishes are abrasive particles in a liquid that is often formulated to give a synergy effect to speed up polishing. Toothpaste is calcium carbonate (ground-up marble) in a gel that abrades teeth surfaces to remove debris and only a little enamel. Some toothpastes may contain corrodents to speed the process. There are countless examples of intentional slurry erosion used in manufacturing processes, but there are also countless examples of slurry erosion as a limiting factor in a process.

- Mining gold from sands
- Handling magnetic media, inks, pigments
- Pumping ceramic slurries
- Water turbines
- Sewage pumps
- Mud pumps

The last slurry problem in this list is worldwide and serious. This mud is pumped downhole in oil wells to cool and lubricate the drill head and to carry the drilling debris to the surface. This mud must be pumped, conveyed, and handled in many ways, and it erodes whenever it contacts solids.

Mechanism

Slurry erosion is sometimes considered to be a type of abrasion. It rightly belongs in the erosion category since it is material removal from a solid surface by the mechanical action of a fluid. The fluid is what applies to force to the solids in a slurry to contact surfaces and produce material removal. Without hard particles, a slurry can be innocuous. It could cause liquid erosion, but liquid erosion is slow unless made energetic by high velocity. Bean soup is a slurry containing nonabrasive particles (beans). Its erosion rate to vessels used in its creation and handling is negligible, unless tribocorrosion comes into play. Suppose the beams raised the pH of the water and made it attack the aluminum pot. We would end up with a situation similar to that in CMP (Figure 8.7). The beans have mass, and when they are stirred against the corroding pot, they can remove the reaction film that suppresses further corrosion, and corrosion becomes more aggressive because now more pot must dissolve for the reaction film to reform. We have tribocorrosion synergy. The corrosion rate without beans rubbing may be 1 µm/h with stirring, and with beans rubbing the rate may be 5 µm/h.

Using the same pot, and the same soup, if we added a half cup of 50- to 70-mesh sand and aggressively stirred with the backside of the wooden spoon, we would add slurry abrasion to our bean soup tribosystem. As shown in Figure 8.9, the beans continue to remove the chemical reaction film, but now sand particles are rubbing on the inside of the pot, and they can remove material without corrosion assist. They can scratch the pot and produce chips in the process of the sand grains rubbing on the pot with a lesser force, and pot material can be transferred to the grains of sand. Thus, we have rubbing-assisted corrosion conjoint with abrasion—material removed from the action of hard substances imposed and moving along a softer substance. Sand-bean soup is not commonly encountered, but sewage sludge and industrial wastewater are not unlike our sand-bean soup.

Slurry pipelines experience material removal like condition b in Figure 8.9, with the exception that there is no spoon forcing the hard substances against the inside of the pipe. The solids are carried by the flowing liquid in the pipe, and they do not rub that much except at direction changes, so the abrasion component of the slurry erosive will be low, but the corrosion synergy may be high. The lightly rubbing solids easily remove reaction

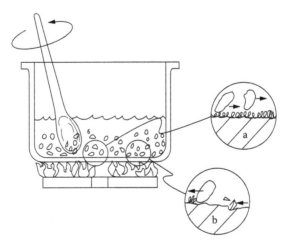

FIGURE 8.9
Slurry erosion from nonabrasive material (beans) and abrasive material (sand).

films (passive surface films), and a corrosion synergy occurs, as in the care of the beans removing the soft film.

CMP is on the other extreme of force of hard particles on the surface. In this instance, there is a rather stiff elastomer pad forcing abrasive particles from the slurry against the rubbing surface. The particles abrade (by scratching and adhesive transfer). In this case, we also have component corrosion synergy.

What is the role of force? Force is a significant factor on any wear process. We get the most force on slurry particles when they are imposed on a surface with some device or action. Pads can be used to rub slurries against a surface, and the force on the particles becomes a function of the stiffness of the pad. For example, a steel pad will produce more force than a rubber pad. A 90° bend in a slurry pipeline will greatly increase the force impacting the slurry particles on the pipe wall. That is why direction changes in pipelines often are addressed by replaceable wear-backs in high-impingement areas.

What is the role of particle size? Intuitively, one would think that when slurry particles get very small, they may become innocuous. However, industrial experience and laboratory tests have confirmed that slurry particles in the size range of 5 to 10 nm can produce a slurry abrasion component. If they can abrade, they can also remove protective films and cause a corrosion synergy. So, even nanometer-sized particles in a slurry can make the slurry very erosive.

What about volume fraction? The volume fraction in the nanoparticle experiments was only 0.2 to 0.3% by weight. So, it does not take very many hard particles to create an erosive slurry.

In summary, the mechanism of slurry erosion is particle removal of protective films from the surface coupled with abrasion from particle rubbing (scratching + adhesive wear). If the slurry is used on a material like plastic or ceramics that do not rely on passive films for corrosion protection, there will only be the abrasion component. Tribocorrosion synergy will not play a role. Finally, the force that imposes the slurry particles on sliding along a surface is a very significant factor. Slurry velocity often provides the particle force, so it is important probably to an exponent of more than 2.

Slurry Abrasivity

What is the role of particle hardness? Particle hardness does make a difference. The harder the particles, the more likely they will scratch a substance. Earlier (Figure 4.1), we presented the Mohs hardness scale. It ranks common materials on their ability to scratch each other. Diamond will scratch all of the minerals on the list, sapphire (aluminum oxide) will scratch everything but diamond and talc, and items on the bottom of the list will not scratch anything on the list. In the 1980s in the United States, a test was developed to rank water slurries of minerals. The test was intended to give guidance on suitable pipeline materials for conveying coal and other minerals long distances (hundreds of miles) in pipelines. If you want to know if, for example, molybdenum ore from XYZ mine can be pipeline transported in a 10% slurry of 500 μm particles in low-carbon steel pipe, this slurry abrasivity test will answer that question.

This test is now an ASTM standard (ASTM G 75), and the test metric is a Miller number. This is a number from 1 to about 1000, with 1000 as the higher abrasivity. The test uses a slurry of interest against a standard reference material (27% Cr white iron). The test rig is

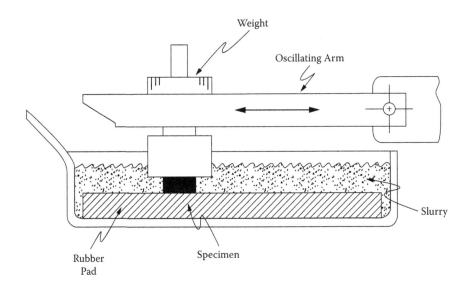

FIGURE 8.10
Schematic of ASTM G 75 slurry abrasion test and Miller numbers for various slurries.

illustrated in Figure 8.10. The slurry is contained in a tray with a rubber lap affixed to the bottom. A dead-weight-loaded arm carries a rider of the reference material (27% Cr iron), and it is reciprocated for 6 hours under standard conditions and the mass loss on the reference iron is measured. Angle plates cause the rider to break contact at the end of each forward stroke to allow the slurry to flow into the rider-rubber contact.

This test also can provide information on possible corrosion synergy. The slurry is made with ordinary distilled water, but if the solids (the molybdenum ore) put in the water change the water's corrositivity, the test will show this. After doing three replicates in the slurry as received, it is buffered to be noncorroding. The mass loss differences with and without buffering determine the tribocorrosion synergy.

This test has been used for about three decades to rank the abrasivity of water slurries containing different minerals. Miller numbers for a wide variety of minerals are tabulated in Table 8.1. This test can be run with riders of different materials to assess how different pipeline materials react to a particular slurry. Slurry pots can also be used to investigate corrosion synergy (ASTM G 119) and do material studies. In summary, slurry erosion looks polished (Figure 8.3) or scratched (Figure 8.4)—all edges get rounded and the surfaces often get shiny. This suggests that material removal is mostly by corrosion or by the adhesive wear component of abrasion. There are tests to study the abrasivity of slurries, and these tests can be used to rank engineering materials for use in slurries.

Liquid Impingement Erosion

Liquid erosion is slurry erosion without particles. Just the mechanical action of an impinging stream can remove protective films and prevent their reformation and create significant local material removal by corrosion, fracture, or physical removal of material. Droplets of liquid can do the same things, and we discussed droplet erosion as a special case since

TABLE 8.1

Miller Numbers for Different Substances

Substance	Miller Number (the higher the number, the more abrasive)
Fly ash	83
Aluminum oxide	1058
Bauxite	70
Calcium carbonate	14
Silicon carbide	1285
Clay	34
Coal	30
Blast furnace dust	57
Gypsum	41
Limestone	36
Magnetite	82
Drilling mud	10
Potash	10
Pyrite	194
Rutile	10
Sand	10
Sewage	25
Shale	53

sonic and supersonic velocities are often involved. For probably the entire life of the chemical process industry, engineers and maintenance personnel have welded plates over tank areas where a chemical or steam is introduced to a process tank. The incoming liquid stream can mechanically remove passive films on stainless steel and produce accelerated corrosion in this area. Water jets can remove material by fracture. High-pressure water is often used to cut materials like foams, fabrics, plastics, etc. An impinging water jet will not cut steel, but it will cut many materials. The cut edges are sometimes not pretty, but they do separate by fracture. If you hold a paper towel taut and inch it into a stream from a faucet, you will see impingement fracture.

The last example in Figure 8.11 is physical removal of material without chemical interaction. The impinging water simply washes away material. It is this mechanism that rivers use to widen and deepen their banks.

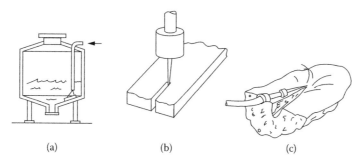

(a)　　　　　(b)　　　　　(c)

FIGURE 8.11

Examples of liquid erosion: (a) steam or high-velocity introduction of a liquid into a vessel; (b) water jet cutting of foam, etc.; (c) physical removal of material by liquid impingement.

Cavitation Erosion

Handling fast-moving water in dams and hydroelectric plants often leads to erosion of concrete structures. Often, the liquid erosion is coupled with cavitation erosion, which is further described in the Chapter 10 on liquid droplet erosion. Imploding bubbles create liquid jets similar to water jet-cutting action and remove material by cavitation erosion (Figure 8.12). Cavitation erosion almost always occurs on tanks that use attached ultrasonic transducers to produce mixing or degassing in a tank. Cavitation on the inside of the tank forms patterns that mirror the vibrational amplitude nodes from the transducers (Figure 8.13). Figure 8.14 shows cavitation damage in a pump impeller handling water.

The damage can be fracture conjoint with removal of passive films. However, very often, the jets are energetic enough to fracture grains from metals or aggregate from concrete. Once the surface is penetrated by a cavitation jet, the start hole easily grows into a

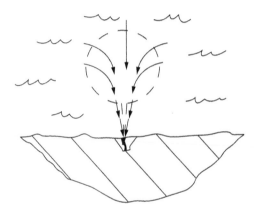

FIGURE 8.12
The mechanism of cavitation erosion. When a bubble collapses, liquid rushes in to fill the void, creating a high-pressure jet.

FIGURE 8.13
Cavitation erosion on a stainless steel tank with ultrasonic transducers on the inside to debubble chemicals. Damage occurs at sound nodes.

FIGURE 8.14
Cavitation damage (bright spots) on a bronze pump impeller.

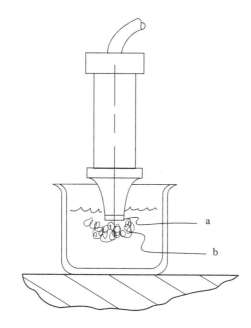

FIGURE 8.15
Schematic of the ASTM G 32 vibrating horn cavitation test. (a) = test specimen, (b) = cavitation field.

macroscipic feature. Bubble collapse jet pressures can be 36,000 psi (248 MPa), comparable to pressures used on water jet cutting. There are two ASTM standard tests for assessing the cavitation resistance of materials: G 32 and G 134. The G 32 test uses an ultrasonic horn to create a cavitation field on the specimen that is affixed to the end of an ultrasonic horn (Figure 8.15). The ASTM G 134 test essentially uses a submerged water jet nozzle to create a cavitation field on a test specimen. Both tests measure cavitation erosion rates and the incubation periods associated with them. There usually is an incubation period that corresponds to initial grain removal or removal of passive films.

Summary

Tribocorrosion has many aspects and many appearances. It means that corrosion can be conjoint with other wear processes like adhesive or abrasive wear, and that the corrosion can be exacerbated by the rubbing process. Slurry erosion mostly produces polished/rounded surfaces like Figure 8.3. Liquid erosion usually polishes metals and fractures particles from concrete and the like, and widens streams by physical removal of material. Cavitation produces pits by fracture (Figure 8.13). Rubbing immersed in a corrosive media allows removal of passive films, for example, in rubbing metal-to-metal hip joints, causing accelerated corrosion and possible particle release that the body must accommodate. Thus, whenever rubbing occurs immersed in a liquid, tribocorrosion may play a role. Whenever liquid velocities are high enough to cause significant forces on a surface, tribocorrosion may occur. The mechanical action of the fluid substitutes for physical rubbing.

Related Reading

ASTM G 5, *Standard Test Method for Making Potentiostatic and Potentiodynamic Anodic Polarization Measurements*, West Conshohocken, PA: ASTM International.

ASTM G 81, *Standard Test Method for Determination of Slurry Abrasivity (Miller Number) and Slurry Abrasion Response/Materials (SAR Number)*, West Conshohocken, PA: ASTM International.

ASTM G 102, *Standard Test Method for Calculation of Corrosion Rates and Related Information from Electrochemical Measurements in Corrosion Testing*, West Conshohocken, PA: ASTM International.

ASTM G 105, *Standard Test Method for Conducting Wet Sand/Rubber Wheel Abrasion Tests*, West Conshohocken, PA: ASTM International.

ASTM G 119, *Guide for Determining Synergism between Wear and Corrosion*, West Conshohocken, PA: ASTM International.

ASTM G 204, *Standard Test Method for Damage to Contacting Solid Surfaces under Fretting Conditions*, West Conshohocken, PA: ASTM International.

Blau, P.J., Celis, J.P., Drees, D., Eds., Tribo-Corrosion: Research, Testing, and Applications, STP 1563, West Conshohocken, PA: ASTM International, 2013.

Brown, S.R., *Materials Evaluation under Fretting Conditions*, Standard Test Procedure, West Conshohocken, PA: ASTM International, 1982.

Miller, J.E., *The Reciprocating Pump, Theory Design and Use*, New York: John Wiley, 1987.

Miller, J.E., Schmidt, F., *Slurry Erosion: Uses, Applications, and Test Methods*, STP 946, West Conshohocken, PA: ASTM International, 1984.

Stansbury, E.E., Buchanan, R.A., *Fundamentals of Electrochemical Corrosion*, Materials Park, OH: ASM International.

Waterhouse, R.B., *Fretting Corrosion*, Oxford: Pergamon Press, 1972.

Young, R.R., *Cavitation*, London: McGraw Hill, 1989.

9

Solid Particle Erosion

Solid particle erosion is progressive loss of material from a solid surface by repeated impacts from fluid-borne solid substances—particles. The particles can range from oxygen molecules, as in outer space, to sand particles in ambient air (Figure 9.1) to roadway stones a centimeter in diameter that produce chips on automobile body parts. The potential for an impacting particle to cause damage to a solid surface depends on its impact energy, specifically its mass (m) and velocity (v). The classic physics relationship for momentum of an object in motion applies: $\frac{1}{2} mv^2$. However, the velocity exponent may be 2.2 or higher.

$$W_s \approx kmv^{2-3}$$

where:
  k = system constant
  m = particle mass
  v = particle velocity
  W_s = erosion rate

The system factor can take into account things like particle shape, roughness, impingement angle, number of restrikes, etc. Figure 9.2 shows severe solid particle erosion (SPE) in a particle conveyance part where the mass of the erodent was large and the velocity was high. A classic experiment was conducted by somebody who never seems to be recognized (including in this work), and he or she showed that the effect of impingement angle on material removal is different for ductile and brittle materials (Figure 9.3). Ductile metals allegedly see reduced erosion at normal impingement because the particles tend to embed and thus do not remove material (Figure 9.4). Conversely, brittle materials break easily on normal impact. A brick mason cuts a brick by repeated impacts normal to the surface piece of the brick spall, and he or she can impact cut a groove that can be hit hard to cut the brick in a single blow. Striking the brick with the brick hammer on an angle only wears the steel brick hammer. The brick wins (Figure 9.5). However, repeated strikes normal to the surface in the same spot dig a big hole by spalling, and this technique is what the mason uses to make a brick cut.

Mechanism

The material removal mechanism in solid particle erosion is fracture—the same as all wear processes that are not conjoint with corrosion where dissolution can be the material removal mechanism. Figure 9.6 illustrates how solid particle impacts remove material. The particles rotate, probably because most are asymmetrical, so the fluid forces are more

FIGURE 9.1
Mild solid particle damage (from particles in air) on the leading airfoil of a jet engine.

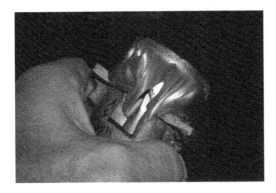

FIGURE 9.2
Severe solid particle erosion of a hose fitting conveying 120-grit alumina particles.

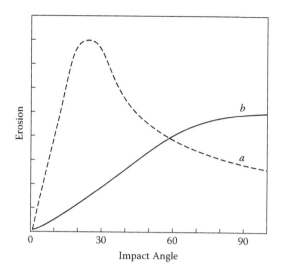

FIGURE 9.3
The generally accepted effect of incidence angle on the rate of solid particle erosion. The dashed curve represents ductile materials; the dotted line indicates brittle materials.

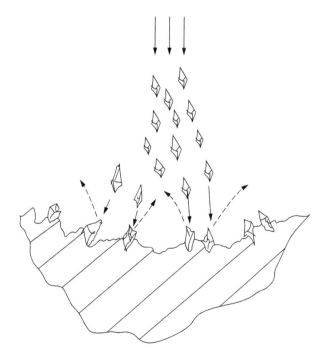

FIGURE 9.4
One possible mechanism for reduced erosion at normal incidence with ductile metals: particles embed and oncoming particles strike themselves.

FIGURE 9.5 (See color insert.)
Effect of impact angle on a brittle material: 20 strikes of a brick hammer at a 20° incidence angle (by thumb) and 20 strikes at a 90° angle to the flat face. Note material removal differences.

on one side than the other. Round particles like glass beads may also rotate, but the damage to the target is likely less than when a particle is angular and sharp (Figure 9.7). The depth of a typical indent is about 1/10 of the particle diameter. A target surface continues to get indented until 100% strike density is achieved (Figure 9.8). At this point, the surface consists of craters, overlapping craters, and shards of material from what were the raised edges of craters. These "extruded" shapes can easily fracture to produce mass loss, and that is how erosion proceeds. Craters form that vary in depth and density based upon the velocity profile of the fluid stream carrying the particles (Figure 9.9). This mechanism of material removal will continue for as long as the particle impingement continues. Most hard coatings will penetrate from solid particle impingement in a concentrated stream (Figure 9.10).

The ASTM G 76 solid particle erosion (SPE) test is a standard test used to rank materials for their relative ability to prevent SPE. It is a pressure blaster with a specific nozzle diameter (1.5 mm), length (75 mm), and nozzle standoff distance (Figure 9.9). The test specifies a particle flow rate of 2 g/mm for 10 min, and this is the test. Mass loss on the test is measured, and it is converted to a wear rate in units of volume per gram of abrasive that impacted the target specimen. Unfortunately, this test is so aggressive that most coated specimens perforate before the designated impingement duration is reached (Figure 9.10). Coatings can be screened by reducing the test time to the point where the coating does not penetrate. The appearance of a test specimen subjected to an ASTM G 76 test is shown in Figure 9.11. The crater profile is like the one shown in Figure 9.12. The crater depth mirrors the velocity profile of the impinging particles. Particles smaller than 10 μm or thereabouts may not have the energy to create significant damage, but above this threshold the damage increases with particle size. Most lab tests use 50 μm or something close in particle size. The aggressiveness varies with the type of abrasive as well. Aluminum oxide is more aggressive than silica and glass beads. Glass beads tend to make surface waves under conditions of saturation impingement (Figure 9.13). The ASTM G 76 test has some problems (too aggressive), but it is the only international test for solid particle impingement.

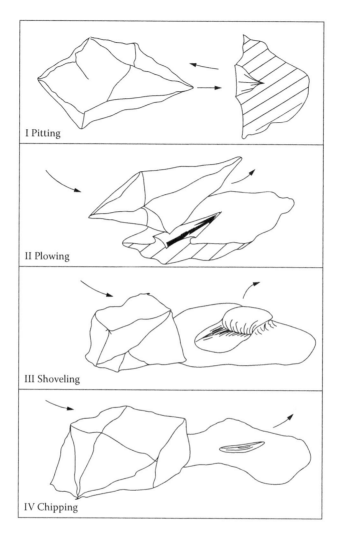

I Pitting

II Plowing

III Shoveling

IV Chipping

FIGURE 9.6
Material removal/displacement mechanisms in abrasive blasting (pressure blast).

Particle Velocity

One of the significant problems in doing solid particle erosion testing is measuring particle velocity. We mentioned that particle velocity controls erosion to at least a power of 2, so how does one know the particle velocity in a pressure blast test rig? Laser Doppler velocimeters are commercially available to do the job, but they are very expensive. A double-disk test rig is what some people use. This is shown schematically in Figure 9.14. The impinging stream of particles is directed at a spinning double-disk assembly. There is a 1 mm wide slot in the first disk and none in the second. The speed of the disk is known,

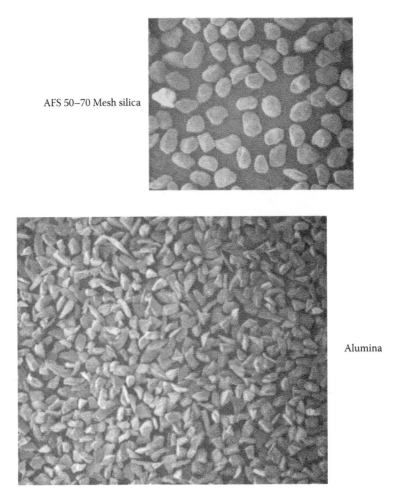

AFS 50–70 Mesh silica

Alumina

FIGURE 9.7
Shape of abrasive particles used in abrasion and erosion studies. The silica (AFS 50/70) is round in shape while the alumina (50 micrometer) is angular.

as is the distance that the particles traveled between the plates. The displacement of the impingement pattern on the second disk from the slot in the front disk is the test metric. Putting this displacement in an equation yields the particle velocity. Other solid particle test rigs are shown in Figure 9.15. There are advantages to each. A pressure blaster is cheap and available, but flow rate and velocity are difficult to accurately control. A wind tunnel gives a lower strike density—more like reality, but it is an expensive rig. The same situation exists for the use of a rocket sled. Particles coming from an accelerator tube provide good velocity control, but there can be particle feeding problems if a gravity feed is part of the rig. Rotating specimens is a reasonable way to get accurate velocity, but the type of rig can become imbalanced if specimens erode differently. Finally, rotating test specimens submerged in particles cannot be done at high velocities because of cavitation, and at slow velocities the particle force is too low to produce measurable damage in a short test time. It may take hundreds of hours to get measurable results.

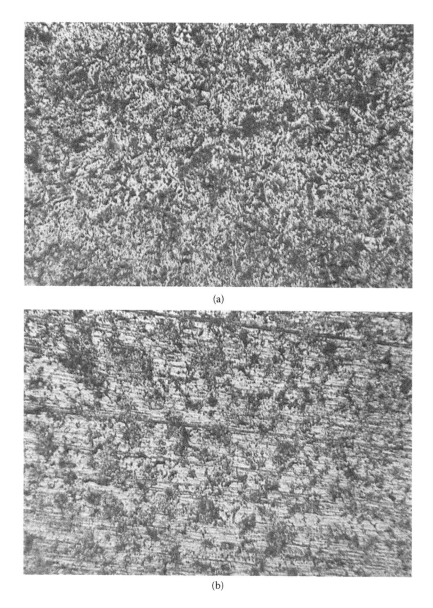

(a)

(b)

FIGURE 9.8
(a) 100% and (b) partial density in abrasive blasting with angular aluminum oxide at 90° incidence (100×).

Manifestations

Simulation of solid particle erosion in the laboratory can produce appearances that run the gamut from gas jet craters to sparsely indented surfaces from wind tunnel testing. The best test to use is the one that best simulates the application where the erosion in taking place.

Figure 9.16 shows a large (4 m diameter) cyclone separator for fly ash that has "wear plates" welded to the vessel to prevent perforation from the impinging particles. It perforated

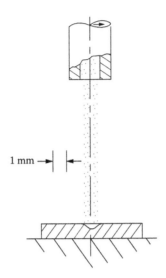

1 mm

FIGURE 9.9
Scale drawing of the nozzle and standoff distance in the ASTM G 76 solid particle erosion test. The test produces a crater with a depth profile that mirrors the velocity of the particles in the impinging stream. Velocity is highest in the center of the stream.

anyways. Figure 9.17 shows a pipeline wear-back that perforated at a change in direction on a pipe carrying powdered coal for a boiler. In both examples, the surface looked like an abrasive-blasted surface in the ASTM G 76 test (Figure 9.11). Thus, the ASTM G 76 gas jet erosion test would be a good lab test to screen materials for this application. Figure 9.2 shows solid particle erosion of a coupling used in conveying aluminum oxide for a surface texturing machine. In this case, the particle stream (220-grit Al_2O_3) was almost parallel to the surface, so the eroded surface was smooth—almost polished. This is what damage looks like on turbine blades in rotors in jet engines that ingest particles. The blades change shape to conform to the aerodynamics of the system, and the surfaces are smooth and shiny, like the sand blast coupling.

In all cases, the mechanism of material removal is fracture. The hard particles fracture particles from the target surface, and the degree of damage depends on the impingement angle, particle size, particle flux, and particle velocity. Soft particles do not erode—they tend to stick to the impacted surfaces. Tiny particles (<10 μm) do not have enough force (mass × acceleration) to do much damage. This form of erosion is often ignored (as in stone chips on vehicles), but it cannot be ignored on steam turbines, jet engines, and exhaust systems where material removal can be great enough to produce failures.

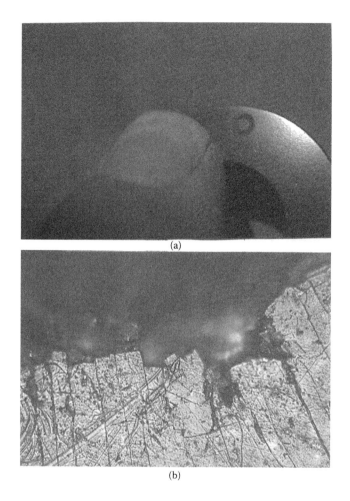

(a)

(b)

FIGURE 9.10
(a) Hard anodize coating after a G 76 erosion test. (b) Note the brittle fracture that occurred at the edge of the crater (50×).

FIGURE 9.11 (See color insert.)
Typical crater developed in the ASTM G 76 SPE test.

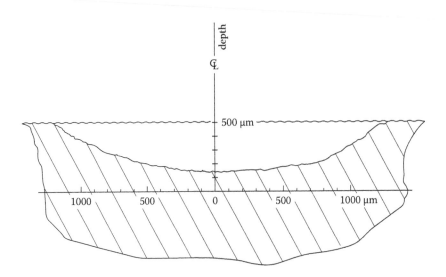

FIGURE 9.12
Erosion crater from G 76 test conducted with low incidence angle (30°) and glass beads as the medium.

FIGURE 9.13
Surface waves produced in 1020 steel from low-angle (30°) impact by 50 μm glass beads 100×.

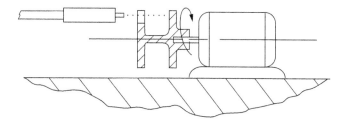

FIGURE 9.14
Schematic of double-disk particle velocity measuring device.

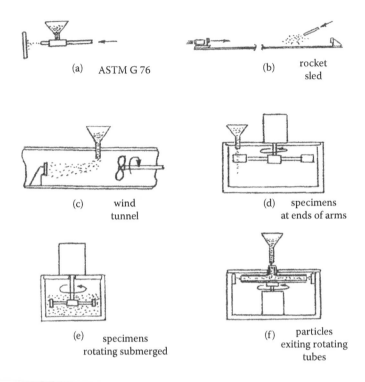

FIGURE 9.15
Schematics of various SPE tests: (a) = ASTM G 76, (b) = rocket sled, (c) = wind tunnel, (d) = specimens at the end of rotating arms, (e) = specimens rotating submerged, (f) = particle's exit rotating tubes.

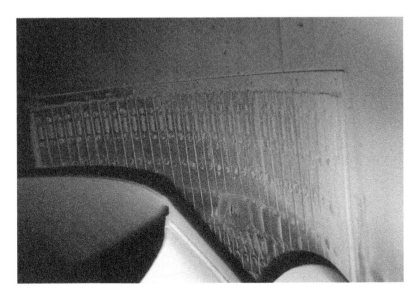

FIGURE 9.16
Solid particle erosion of a large cyclone separator handling boiler ash. The wear plates perforated, as did the vessel wall.

FIGURE 9.17
Solid particle erosion of a pipeline elbow wear-back. The pipeline was carrying powdered coal.

Related Reading

ASTM G 76, *Standard Test Method for Conducting Erosion Tests by Solid Particle Impingement Using Gas Jets*, West Conshohocken, PA: ASTM International.

Bachelor, W., McGee, J., *Wear Processes in Manufacturing*, STP 1362, West Conshohocken, PA: ASTM International, 1998.

Bahadur, S., Ed., *Effect of Surface Coatings and Treatments on Wear*, STP 1278, West Conshohocken, PA: ASTM International, 1996.

Denton, R., Keshovan, M.K., *Wear and Friction of Elastomers*, STP 1145, West Conshohocken, PA: ASTM International, 1992.

Huchings, I.M., *Monograph on the Erosion of Materials by Solid Particle Impact*, MTI Publication 10, Columbus, OH: Materials Technology Institute, 1983.

Raask, E., *Erosion Wear in Coal Utilization*, New York: Hemisphere Publishing, 1988.

Stachowich, G., Batchelor, A.W. *Engineering Tribology*, 3rd ed., Amsterdam: Elsevier, 2004.

Strafford, K.N., Datta, P.K., Googar, E.G., *Coatings and Surface Treatments for Corrosion and Wear Resistance*, Birmingham: Institution of Corrosion Science & Technology, 1984.

10

Liquid Droplet Erosion

Droplet erosion is progressive loss of material from a solid surface caused by the action of impinging liquid droplets. It is similar to solid particle erosion, only droplets of liquid are impinging on the surface rather than solid particles. The mass (size) of the droplets and their velocity are important factors in determining the severity of the damage. This mode of erosion usually only occurs at droplet velocities greater than 50 m/s (100 mph). Cars and trucks do not get damaged by raindrops because the velocity is insufficient to damage paint on steel, plastics, glass windscreens, and the other things that protrude from a car or truck. Subsonic aircraft (<400 m/s) can be damaged by flying through rain fields, but exposure is usually limited to takeoff and landings. The number of droplet strikes is also a significant factor in determining the severity of damage—more droplets impinging, more potential for damage.

Droplet erosion is costly in steam turbines where condensate droplets impinge on rotor vanes in the cooler sections of the compressor. The vanes are "machined" by particles so that their shapes mimic the velocity/flux variations in the droplet stream. Wind turbines will supply a significant portion of the energy used in many countries. Some have goals of 20% by 2020. The huge blades on these turbines, of course, will run in rain, snow, ice, dust storms, etc. The top speed on a 100 m diameter wind turbine can reach velocities of 100 m/s, and thus the materials that they are made from will be subject to droplet erosion. They are sometimes made from composites that are particularly prone to droplet and particle erosion—mostly because UV from sunlight degrades the polymer matrix and droplets "seek out" surface weaknesses.

Airplane propellers, helicopter rotors, and jet engines are subject to droplet erosion. Planes fly through rain fields. The vertical takeoff planes used in the military often take off from ships, and in these cases, their engines can "ingest" seawater droplets at speeds capable of producing liquid droplet erosion (LDE). Thus, there are many applications where droplet erosion can be a significant, if not limiting, factor, and this form of erosion must be dealt with. We will show some examples of what it looks like and describe the testing techniques that are used to simulate droplet erosion to screen candidate materials for use in droplet erosion conditions.

Droplet Damage

Figure 10.1 is a simple but classic example of droplet erosion. The droplets fall from the roof of a tall building (church) directly onto a concrete sidewalk. In this case, the velocity is small, maybe 10 m/s, but the duration is large—the droplets have been falling on this sidewalk since 1964 (48 years), and the concrete matrix has been eroded one sand particle at a time, leaving the concrete aggregate standing proud. This is not a life-threatening

(a)

(b)

FIGURE 10.1 (See color insert.)
(a) Roof of a building; (b) walkway from a distance; (c) close-up of droplet erosion from the walkway.

situation. The concrete sidewalk just gets rougher and nobody pays much attention, but this illustration shows how droplet erosion works.

The mechanism of material removal is cracking or fracture of the weakest areas on a surface and propagation of those cracks to produce material removal. The force that produces the cracking is that of the impinging droplets. Figure 10.2 shows how droplets seek out surface weaknesses. Figure 10.3 shows how forces are distributed in a surface when a

(c)

FIGURE 10.1 (continued) (See color insert.)
(a) Roof of a building; (b) walkway from a distance; (c) close-up of droplet erosion from the walkway.

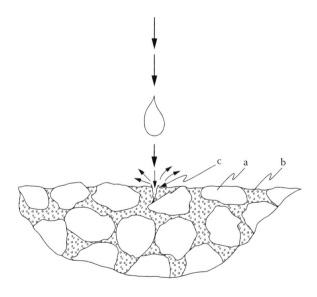

FIGURE 10.2
The mechanism of the concrete erosion cited in Figure 10.1. Droplets strike the concrete with a force proportional to their mass and velocity. These forces initiate cracks in the cement phase of the concrete and sand particles erode out of the cement, leaving aggregate standing proud. Aggregate will eventually fall out as cement erosion progresses.

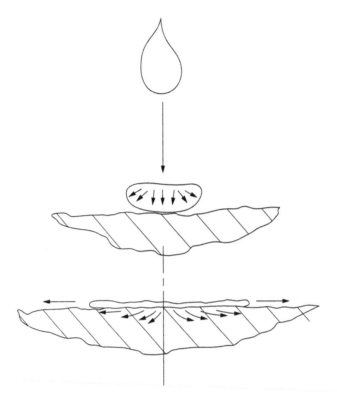

FIGURE 10.3
Compression of a droplet on impact and the radial forces produced from that compression.

droplet hits and forms a splat. Figure 10.4 shows the various ways that damage initiates. An impinging liquid droplet will produce force relating to its mass and impact velocity. As is the case in solid particle erosion, the droplets must have a sufficient mass for the force to be significant. Droplets that damage most materials have diameters in excess of 50 μm. The droplet velocity has to be significant as well—usually greater than 100 m/s, and the effect of velocity is raised to a power. Thus, the forces can be significant and they are substantial over the splat size. Weaknesses in a solid surface can be small, like 50 μm, so when these energetic droplets hit a weakness in a material like a grain boundary, they can fracture small amounts of material. With repeated impacts, the damaged spots get deeper and wider, and soon grains can be fractures or microconstituents can be fractured from the surface. In the case of our sidewalk example, aggregate particles will pop out and the droplets will start to erode the next layer down, and on and on.

Coatings almost always have defects/discontinuities that can initiate droplet damage. Paints on airplanes have thickness variations, tiny bubbles, included particles, etc. These will be potential sites for droplet erosion. It is common for there to be an incubation period wherein these small cracks and fractures start to get large enough for major fracture. Figure 10.5 shows typical mass loss progression in a typical laboratory droplet erosion test. There is an incubation period followed by a steady-state erosion regime. Droplet erosion is not unlike cavitation erosion; both forms of erosion are produced by high-velocity impingement of a liquid. Sometimes cavitation erosion testing is used to screen candidate materials for liquid droplet erosion application because cavitation testing is usually simpler and less costly. Figure 10.6 shows cavitation damage around fluid passages in a stainless steel plate.

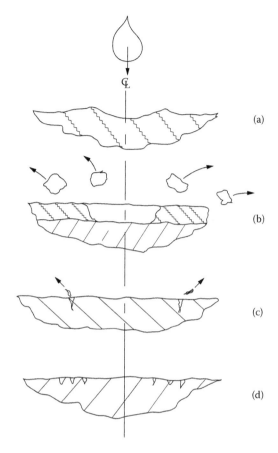

FIGURE 10.4
Damage modes on droplet impact: (a) = plastic deformation, (b) = spalling, (c) = cracking, (d) = pitting. These can be conjoint.

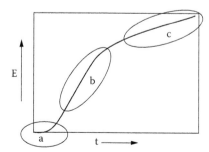

FIGURE 10.5
A typical volume loss (E) versus time (t) plot from a laboratory droplet erosion test (ASTM G 73) showing incubation (a), maximum erosion rate (b), and terminal erosion (c).

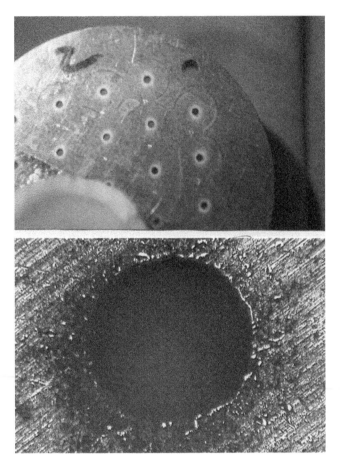

FIGURE 10.6
Cavitation damage on the exit of a diffuser plate. Cavitation is similar to liquid droplet erosion in that material removal comes from liquid impingement. The damage is shown at 100× magnification.

LDE Testing

The ASTM G 73 test is shown schematically in Figure 10.7. There are other rig configurations that will comply with this test method (Figure 10.8), but basically, the test uses test specimens mounted on the periphery of a rotating arm or disk, and at least once per revolution, the test specimen strikes a liquid droplet field and the droplets are usually smaller (50 to 100 μm in diameter) than the test specimen. The specimen velocity as it strikes the liquid droplet field will depend on the tangential velocity of the specimen affixed to a disk or arm. The test method cites speeds. The standard practice uses droplet velocities in the range of 18–600 m/s. The standard is a practice rather than a test method, and this allows the use of different testing and test procedures. Test specimens can be rounds, flats, or other shapes, and the usual erosion measurement is mass loss (converted to volume loss) over a certain time interval. The test practice requires the development of a cumulative erosion curve like that shown in Figure 10.5. The practice gives an empirical relationship for incubation time to illustrate the key factors in determining the onset of significant erosion damage:

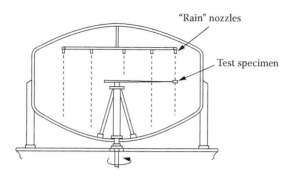

FIGURE 10.7
Schematic of the ASTM G 73 test.

FIGURE 10.8
A horizontal rotating arm liquid droplet erosion test rig with specimens on both ends of a rotating arm and two droplet injectors.

$$t = 10 \, (HV)^2 \, Km \left(F_1(V/100)^{4.9} \right)$$

where
t   = the incubation time(s)
HV = Vickers hardness number
Km = a material parameter
F_1 = specific impact frequency
V   = impact velocity m/s

This formula demonstrates that hardness is a key material property in controlling droplet erosion, velocity, and frequency of impact, and are also significant factors. These tests are usually run with a reference material of known erosion resistance. For example, for aircraft applications, the reference material may be the aluminum used in the leading edges of wings, the clear plastic used for windows, or the paint used on the nose of the plane. Clear plastics may be evaluated by haze—the effect of the surface damage on the light transmittance of the plastic. Aircraft windows are often damaged by droplet erosion. Paint damage on aircraft often occurs on leading edges and radomes on the nose of the aircraft. Fiber-reinforced plastics often "shred" under high-velocity droplet impingement conditions. Long-term droplet erosion in a heterogeneous material may look similar to the droplet erosion on concrete shown in Figure 10.1.

Summary

Overall, liquid droplet erosion is very much like solid particle erosion, except that a liquid drop behaves much different than a solid particle on impact. The solid particle will rebound, dent, or cut the target surface. A droplet splats, and the force profile will likely have a Gaussian form, with the highest force coinciding with the center if a droplet. The splat produces a tangential force that can also be very energetic. Liquid droplets in the form of rain do not damage most engineering materials because the velocity is produced by gravity and is likely to be below 10 m/s in velocity. However, if the frequency (number of drops impacting) becomes very large, as was the case in the concrete below the roof eaves in Figure 10.1, then the erosion can become significant.

LDE is a special form of erosion that becomes a limiting factor when velocities get very high. Turbine devices are very susceptible to damage (Figure 10.9), as are all aircraft,

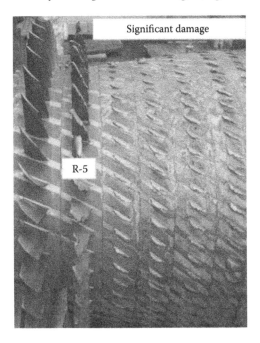

FIGURE 10.9
Liquid droplet erosion on blades in the compressor section of a steam turbine. (Courtesy of David Gandy, Electric Power Research Institute (EPRI).)

satellites, etc., that must fly through rain fields. Thus, it is a very serious form of erosion, but only to selected applications.

Related Reading

Adler, W.F., Ed., *Erosion: Prevention and Useful Applications*, STP 664, West Conshohocken, PA: ASTM International, 1979.

ASTM G 32, *Standard Test Method for Cavitation Erosion Using Vibrating Apparatus*, West Conshohocken, PA: ASTM International.

ASTM G 73, *Standard Test Method for Liquid Impingement Erosion Using Rotating Apparatus*, West Conshohocken, PA: ASTM International.

ASTM G 134, *Standard Test Method for Erosion of Solid Material by Cavitating Liquid Jet*, West Conshohocken, PA: ASTM International.

Blau, P.J., Ed., *Friction, Lubrication and Wear Technology*, Vol. 18, Materials Park, OH: ASM International, 1992.

Davis, J.R., Ed., *Nickel Cobalt and Their Alloys*, Materials Park, OH: ASM International, 2000.

Erosion by Cavitation or Impingement, STP 408, West Conshohocken, PA: ASTM International, 1968.

Evaluation of Wear Testing, STP 446, West Conshohocken, PA: ASTM International, 1968.

Preece, C.M., Ed., *Treatise on Material Science and Technology in Erosion*, Vol. 16, San Diego: Academic Press, 1979.

Trevina, D.H., *Cavitation and Tension in Liquids*, Philadelphia: Adam Hilger, 1987.

11

Sliding Friction

The ASTM G 2 Committee on Wear and Erosion maintains a list of definitions for friction, wear, and erosion terms: ASTM G 40. Unfortunately, this list of terms does not include *friction*. It defines friction-related terms: friction force, coefficient of friction, kinetic coefficient of friction, static coefficient of friction, and stiction. The dictionary definitions of friction range from "rubbing" to "the resistance that a body encounters when movement on another body is attempted or sustained." The terms *rubbing* and *resistance* are included in the majority of the dictionary definitions. However, friction is encountered in fluids even between molecules rubbing in a solid. For example, plastics can locally heat when they are repeatedly stressed as in a fatigue test. Also, it is well known that friction of ship hulls versus water is a significant factor for fuel usage. A proper definition of friction should include all of these types of friction, all of those shown in Figure 11.1:

Sliding—solids rubbing

Rolling—a solid rolling on a solid surface

Lubricated—sliding or rolling with lubricant in the contact

Internal—within a solid

Atomic—between the atoms of contacting surfaces

An inclusive definition might be the force resisting the start of motion or continued motion of one body or substance on or within another body or substance. The problem with this definition is that friction is also inextricably connected with energy dissipation. As mentioned, some materials like automobile tires get hot from flexing. This heat must be dissipated. A significant amount of energy, heat, must be dissipated in braking a vehicle or moving part on a machine. Thus, rubbing solids generate heat. Forces resisting motion generate heat. And this is logical since the forces occur in sliding and sliding involves relative motion or distance, and we know that a force times a distance yields an energy term. However, attempts to add an energy term to a friction definition have been met with dissension among the tribology researchers. So, suffice it to say at this point that friction is a resisting force that must be overcome to start or to sustain motion.

Types of Friction

Just as there are modes of wear, there are types of friction. Sliding friction occurs when solids slide on each other with the contacting surfaces unchanging. If a box is placed on the floor, one face of the box always contacts the floor. If the box is small and the floor is extremely rough, it may start to tumble on different faces and rolling friction would be the mode of friction predominating. Rolling occurs on shapes other than spheres and

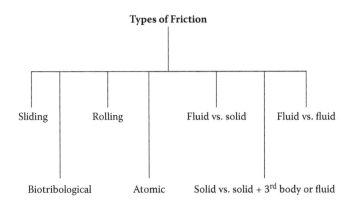

FIGURE 11.1
Types of friction.

cylinders. In fact, in abrasion situations, particles often abrade surfaces by rolling contact. Now back to the box on the floor example. If the floor is smooth but contaminated with sand or dirt particles, pushing the box will be easier than pushing it on a clean floor. It will be supported by particles that will roll and support the box. Sand and dirt particles are not normally spherical; they have many shapes, mostly angular, but under these conditions they will roll and rolling friction usually will be less than sliding friction. Rolling friction is a resistance to rolling. We will dedicate Chapter 12 to it, so suffice it to say here that rolling and rolling friction can occur with objects that normally do not roll. The box on the floor will roll if, for example, it was placed on an inclined plane and the plane was continually raised. There would be an inclination angle at which the box would roll down the plane (or fall on the floor).

Lubricated friction is a special type of friction because the friction force manifested will depend on the thickness of separating lubricant film. We discussed the Stribeck curve in the chapter on lubricated wear (Chapter 7). This iconic curve shows that the system friction is high when solid-on-solid is likely and lower when the oil film starts to completely separate the solids in relative motion. The nature of the separating lubricant film is all-important.

Internal friction is something that has been studied for decades in metals, but it probably can be important in other material systems. Internal friction in pure iron is the damping of vibration-induced stresses by atomic activity within the metal. The classic test for measuring internal friction is to make a torsional pendulum of the material of interest. The pendulum is set into motion and the damping of successive vibrations is measured by the logarithmic decrement—the natural log of the amplitude of successive vibrations.

$$Q = \ln k / N \pi$$

where:
  Q = measure of internal friction
  k = amplitude decay in one cycle
  N = total number of vibration cycles

Q is a parameter that can be used to assess the damping capacity of iron and other materials. It has application in ranking the ability of the material to damp noise or other kinds of vibrations. However, the concept that is germane to this discussion is internal friction. There is resistance to continued motion of torsional vibrations caused in the case of the pure iron, by carbon atoms migrating between lattice sites, and the more carbon atoms, the more internal friction—the more damping. In long-chain plastics and elastomers, internal friction may arise from polymer chains slipping with respect to each other. In polymer composites, internal friction may arise from interaction between fiber reinforcements and the polymer matrix. In ceramics, internal friction may arise from atomic or crystal changes produced by motion-induced strains.

Internal friction in pure iron is a tool to measure carbon content. In polymers, internal friction can cause heating in cyclic stressing, and this heat can initiate property changes, which can be detrimental. Thus, internal friction is resistance to motion produced by internal reactions rather than external rubbing, and it can be a factor in some tribosystems.

Atomic friction is resistance to motion caused by energy barriers and interactions produced when atoms of one body slide in atomic contact with atoms from another body or substance. Molecular dynamics is widely used to study what happens when, for example, a spherical gold rider consisting of 100 atoms slides on a flat copper surface of 1000 atoms. Computers model the forces and atom migrations. Atomic friction has become important in nanodevices. Normally, functional surfaces are covered with many types of contaminants, films, oxides, and the like so that when a macroscopic piece of gold slides on copper, the gold atoms never really contact the copper atoms. There are many atomic layers of "trash" between them. However, nanotechnology is creating devices that may be only 10 atoms by 20 atoms in size, and, under these conditions, atomic friction can be a limiting factor. Atomic force microscope tips can be made to be only a few atoms in radius. When it is used on atomically clean surfaces, it can record atomic friction. In fact, the atomic friction on a 10×10 nm area can be mapped.

There is also the friction that occurs in biotechnology. Skin friction is of great importance in many industries, like the clothing and cosmetics industry. The resistance, the feeling, that a person senses upon stroking an article of clothing or skin often determines if he or she will purchase it. People like a pleasant feel; however, their feel was really a measurement of sliding friction—hands versus fabric or hand versus skin. Modification of hair and skin friction is a billion dollar industry in the form of conditioners, hand creams, soaps, and the like. Medical devices often involve the friction of internal organs versus metal, and these factors have to be dealt with. Catheters need to slide easily into body parts, so it is necessary in designing these devices to measure, for example, the friction of stainless steel and other materials versus the inside of veins and arteries. Then there is the concern of blood flow within implanted devices. There is a need to know the friction of blood versus synthetic arteries and the like. Many biotechnology friction applications involve sliding friction, but because at least one member of the sliding couple has a surface consisting of living cells, the situation is much different than if both sliding members are not living substances. Biotechnology often requires tribotesting in in vivo conditions.

Finally, the friction of liquids versus liquids and solids is a very common occurrence, but one not normally discussed in tribology; the fluid mechanics people try to maintain ownership of this technology. However, U.S. government vehicle fuel economy mandates in 2010 made tribologists take notice of the friction of liquids versus solids in the form of viscous losses in internal combustion engines. Some estimates place viscous losses to be 10% of an engine's power. That is, 10% of an engine's horsepower is consumed by overcoming drag produced by moving parts in and out of contact with oil in the engine sump. The response to viscous drag had been to lower the viscosity of motor oils from 10 W 20 to 5 W 20 and eventually to 0 W 20. Thus, motor oils are becoming water-like in viscosity to lower viscous friction, friction between metals and oil.

There can also be viscous friction effects between liquids having different viscosity. Industrial mixers are used to blend paints, food items, process fluids—countless liquids, with the viscous behavior of the liquids being blended producing the friction effects. This kind of friction can be important in the manufacturing process. It is a lot easier to blend milk uniformly into mashed potatoes than cheese into the same mashed potatoes.

In summary, there are various types of sliding friction. Some have been well studied and characterized, and some have not. This chapter will present examples/manifestations of those types of sliding friction that are most important in design engineering.

Friction is very important. The life of every car and truck driver is dependent on sliding friction in the vehicle's brakes. Vehicle braking systems have become so reliable that most drivers take their brakes for granted. However, there is a great deal of tribology and materials engineering involved in making the systems work as well as they do, but the point is that sliding friction stops the vehicle on demand. How important is that?

Under the hood in automobiles and trucks, rubber belts drive the accessories needed to make the engines work: the radiator fan, the power steering, the water pump, the air conditioner, etc. So friction is very important in power transmission. Sliding friction is controlled by the rubber characteristics and tensions in the belts so that sliding does not occur. The same type of friction control is needed on footwear. Soles of shoes and walking surfaces must be engineered so that heels of footwear do not slip on them and produce falls and related liability. Sliding friction needs to be low in many applications, like conveying items on chutes. Skiing and ice skating require low friction to work. Sailing races are won by boats what hull paints that reduce the viscous effects of water against their surface. Large ships select hull paints that are antifouling to reduce water-hull friction and reduce fuel consumption. Even Olympic swimmers use textured suits to reduce water-body friction. And, of course, friction effects are huge in clothing and furniture. People want fabrics that feel good and slide comfortably on their skin. Furniture makers must rely on leather friction modifiers so that people do not slip out of chairs when they sit down.

Sliding friction is an integral part of everybody's life, but we have learned to control it so well that in most applications it is no longer a concern. However, every time that something new is developed, new sliding friction concerns arise, and this chapter is intended to address the new and old unsolved friction concerns. That is the objective of this chapter: to present accepted practices for measuring, recognizing, and dealing with sliding friction problems.

Knowledge of the friction force is documented in the ancient Greek writings, and the laws of friction have been known for centuries. This chapter will review the historical evolution of friction since these laws are still applicable. Then the fundamentals of sliding friction will be discussed along with the factors that affect sliding friction. Friction force measurement techniques will be discussed as they apply to analyzing friction events in mechanisms and machines. Then the types of friction force output that can be encountered in machines will be demonstrated and discussed. The chapter concludes with a brief discussion dealing with sliding friction-related problems.

Early Studies of Friction

Famous Greek philosophers wrote about friction force in about 330 B.C. However, cave drawings suggest that prehistoric people used logs as rollers to move heavy loads. They learned that rolling friction is less than sliding friction. Wheeled carts were in use around 4000 B.C., and the Egyptians used lubrication to reduce the friction force encountered in moving the massive stones to build their pyramids and other architectural wonders. We

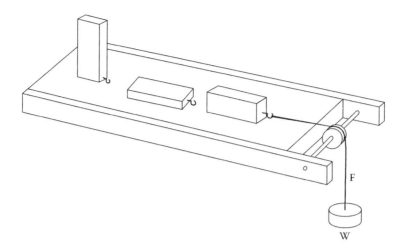

FIGURE 11.2
Leonardo da Vinci's sliding friction experiments.

also know that prehistoric civilizations used the sliding friction of rotating sticks driven by string bows not unlike the bows used for bow and arrows; they rotated a vertical stick rapidly on a flat stick surrounded by tinder to start fires. Thus, use of sliding friction was an essential part of mankind's existence. The Greek and Roman civilizations made great application of wheels and sledges to deal with friction. There is strong evidence to show that animal fats and the like where use as lubricants in Roman and Greek periods. Records are scarce and friction learnings are not well documented from the end of the Roman Empire to the Middle Ages. However, in 1490 or thereabouts, Leonardo da Vinci performed experiments that were the basis of much of what we know about sliding friction today. Being the great artist that he was, he made detailed sketches of his experiments that miraculously were preserved to document what he did. Figure 11.2 shows schematics of da Vinci's famous experiments that produced the first and second laws of sliding friction.

1. The friction force is proportional to the applied load.
2. The friction force is independent of area.

The experiment outlined in Figure 11.2 established the first law. He placed various weights on the sled and measured the force needed to pull the sled. The more weights, the more force needed to pull. He also recorded that the friction force was about 0.25 times the weight of the sled. He established the second law of friction by using oblong blocks like those shown in Figure 11.2. No matter what face he slid the block on, the force to start sliding and sustain motion was the same even though the contact areas were very different.

The French scientist Guillaume Amontons, in 1699 or so, was credited with the mathematical expression for the coefficient of friction:

$$F = k N$$

where:
F = friction force
k = constant (at the time he concluded it was one third for the materials that he used in his studies)
N = normal force pressing the bodies together

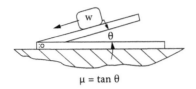

$$\mu = \tan \theta$$

FIGURE 11.3
The coefficient of static friction as the tangent of the angle at which slip occurs on an object on an elevating plane.

In 1733 or thereabouts the mathematician Euler was credited with the use of the now famous Greek symbol μ for the k in the preceding formula. It is the coefficient of friction. In this era it was also learned that the friction force to initiate motion of a body at rest was usually higher than the force required to sustain motion on the body after motion started. Thus, coefficient of friction was recognized as a dimensionless ratio of the friction force to the normal force, F/N. It was given the Greek symbol μ, which is still in use, and early researchers identified a difference between starting friction force, F_s, and friction to sustain motion, F_k. Around the same time, the mathematics were developed to show that the tangent of the angle at which a body on an inclined plane starts to move is equal to the static coefficient of friction (see Figure 11.3):

$$\mu = \tan \theta$$

Also about this time the concept and mathematics were developed for viscosity, which is a form of internal friction.

By the time of the Industrial Revolution, the world had the benefit of many significant learnings regarding friction. Also, by that time, countless experimenters unsuccessfully tried to develop a perpetual motion device by eliminating all friction forces. Even Leonardo da Vinci engineered a device that would pump water continuously, as shown in Figure 11.4; he conquered friction. The device may never have been built, but at any time there are many devices that claim zero friction in their mechanisms. However small it might be, there is always some type of frictional resistance to motion that will eventually

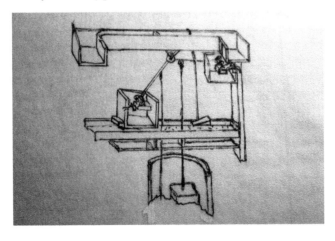

FIGURE 11.4
One of Leonardo da Vinci's machines that would operate continuously—perpetual motion.

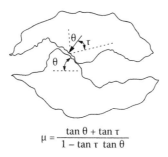

$$\mu = \frac{\tan \theta + \tan \tau}{1 - \tan \tau \tan \theta}$$

FIGURE 11.5
A model for coefficient of friction based upon the rugosities on one surface sliding up the rugosities on the mating surface.

stop a device. There are clocks in museums that will run for 100 years without attention; 100 years does not constitute perpetual motion. Another friction learning from the Middle Ages is the concept that sliding friction of the solid body of another solid body was related to the "rugosities" on their surface, that is, its surface texture and shape features. Like the zero-friction concept, there have been countless studies into the role of surface texture and friction. Even today there are hundreds of mathematical models relating the coefficient of friction to surface texture parameters, such as the one shown in Figure 11.5. We will discuss the role of friction surface texture in detail later in this chapter, but the concept that sliding friction has its origin solely in surface texture interference on the mating surfaces has been disproved countless times. Researchers have used cleaved crystals that are atomically flat and smooth and measured the friction coefficients of these self-mated materials. The sliding coefficient of friction is often very high, and it is never zero.

In the 1950s, one of the foremost tribologists in the United States, Ernest Rabinowitz, proposed the explanation of the area independence of the friction force. He may not have been the originator of the explanation, but he popularized the explanation in his books and tribology courses. The explanation is as follows:

$$F = \mu N$$

and the friction force F arises from the shear strength of surface asperities that need to be sheared in order for motion or friction to occur (Figure 11.6). Asperities a, b, and c are touching and bonds occur between surfaces A and B at these points. Thus, the friction force F is equal to the shear strength, S, of the asperity junctions times the number of junctions (their real area of contact, a):

$$F = S\,a$$

It follows that

$$\mu = S\,a/N$$

Also, the real area of contact, a, is determined by the ratio of the normal force to the penetration hardness, P:

$$a = N/P$$

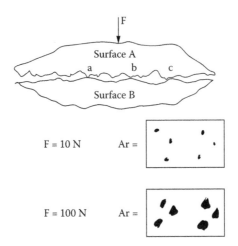

FIGURE 11.6
The concept of surfaces contacting at "asperities" (like a, b, and c) and the real area (Ar) of contact is a function of the force pressing the bodies together.

The applied load is carried by the real area of contact and the size of the contact depends on the penetration hardness, P. Combining these equations:

$$\mu = F/N = Sa/Pa = S/P$$

So, the coefficient of friction is really the ratio of the shear strength of contacting junctions and the penetration hardness of the weaker member of the sliding couple. This explains the area independence. The real area of contact increases as the force pressing the bodies together increases. It also explains why the coefficient of friction for most couples is less than 1. The shear strength of the material and its hardness do not vary by orders of magnitude, only percentages. It also explains why sliding velocity does not have a big effect on the kinetic coefficient of friction of a couple. Strain rate does not have orders of magnitude effects on shear strength and hardness.

By the 19th century most of the fluid friction concepts had been developed, as well as all sorts of formulas for the friction effects of devices like winches, pulleys, blocks, etc. The concept of fluid friction arising from tees, elbows, and related plumbing devices was in place, and hydraulics and fluid mechanics were established to deal with the friction effects of fluids versus solids and the internal friction of liquids: viscosity and resistance to fluid mixing. In 1900 or so, Richard Stribeck and others developed the Stribeck curve (see Chapter 7) that explained friction in fluid-lubricated sliding systems. This landmark work continues to serve as a basis for analysis of lubricated sliding friction.

Overall, much as been learned about friction over the past three millennia, but we are still using friction laws developed in the Middle Ages and the same testing techniques used for hundreds of years. Computer modeling in the form of molecular dynamics has given us insight into the atomic and molecular reactions that occur when atoms of one material rub on another material, but such models do not apply to the complicated surface machines that are covered by oxides, contaminants, water vapor, oils, etc. After the many centuries of studying sliding friction, we still need to simulate tribosystems in the lab to measure friction forces. This chapter will show readers what friction force effects to expect in typical tribosystems.

Fundamentals of Sliding Friction

Friction is a force; dealing with it means external energy will be required to overcome the friction, and that friction energy is dissipated in some fashion. This is how friction is inextricably associated with energy. A body at rest is acted on with the force tending to produce motion; the force will usually initially increase, until macroscopic motion occurs (Figure 11.7). When F_s is reached, motion initiates and a force to keep W in motion lowers to F_k.

F_s is used to calculate the static coefficient of friction.

F_k is used to calculate the kinetic coefficient of friction.

Our first caveat regarding this ideal situation is that for some tribosystems and material couples this ideal scenario does not happen. Plastic–plastic couples sometimes do not show that static force is greater than the kinetic force. The situation may look like Figure 11.8.

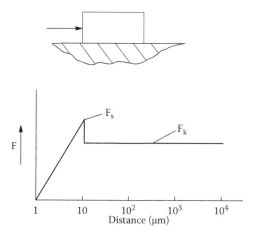

FIGURE 11.7
Friction force at the start of motion.

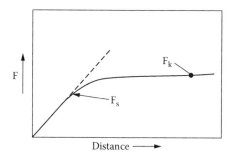

FIGURE 11.8
Friction force at start of motion on a couple that does not have a pronounced breakaway force—use the transition from linearly increasing force.

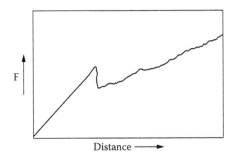

FIGURE 11.9
Friction force never stabilizes; this suggests that the tribosystem is still changing. Force needs to be monitored longer.

In this instance, the point where the force-distance curve departs from linearity is used to calculate the static coefficient of friction, and the kinetic coefficient of friction is calculated from the plateaued force. The second caveat about breakaway friction force is that for some couples and tribosystems there may be a breakaway force spike, but the kinetic force never stabilizes, as shown in Figure 11.9. This kind of force recording suggests that the rubbing services are changing. They may be wearing. When surfaces are wearing, you are not measuring the friction characteristics of a particular mating couple, but the friction of the mating couple separated by wear detritus. For this reason, friction coefficients calculated from wear test data should not be construed as the proper friction coefficient for that couple under different wear circumstances. The next caveat to the perfect situation shown schematically in Figure 11.7 is that movement is really taking place before macroscopic movement is observed. Some people call the motion between the start of macroscopic motion friction creep. Every time a force is applied to a body, something must deflect or strain to accommodate the applied force. It may be plastic deflection of surface features like the rugosities, etc. Some researchers study this phenomenon to try to understand why it occurs. Why is there an initial spike when motion is attempted? Intuitively, elastic deflection of contacts at the true area of contact makes sense. Hooke's law dictates that there is a strain for every applied stress, so there must be some motion, if only elastic deflection of surface features. It is also logical for sliding to require less force than F_s since F_s signals the establishment of new surface contact, and there is no time for accommodation of meeting surface features. From the practical standpoint, most engineering applications are adequately studied by just measuring F_s and F_k and ignoring the motion, deflection that occurs at forces less than F_s.

Next, there is a caveat that applies to F_k. Sometimes stick-slip motion occurs. A friction force recording will look like Figure 11.10. The ASTM G 40 terminology standard defines stick-slip as "a cyclic fluctuation in the magnitudes of friction force and relative velocity between two elements and sliding contact usually associated with a relaxation oscillation dependent on elasticity in the tribosystem characterized by a decrease in the coefficient of friction with onset of sliding or with increase of sliding velocity."

Basically, stick-slip is a phenomenon that is produced by elasticity in the sliding system. If you pull an object with a nylon string, you could get stick-slip behavior. If you pulled the same object on the same counterface with a no-backlash steel lead screw, you would not get stick-slip behavior. However, some tribosystems by their very nature involve motion produced with significant elasticity and, thus, are prone to stick-slip. An example would be sliding plastic films on each other. Both members have a high degree of elasticity. A

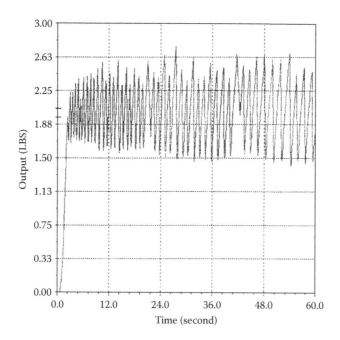

FIGURE 11.10
What stick-slip looks like in a force-distance recording. It is a harmonic vibration.

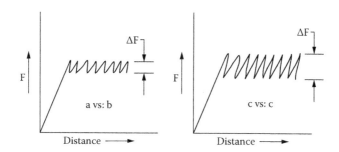

FIGURE 11.11
Comparison of couples that both exhibit stick-slip behavior.

very common manifestation of stick-slip friction is a squeaky door hinge. These are usually unlubricated steel-on-steel couples, and the cantilevered design of the hinge area produces rather elastic conditions. If stick-slip is the friction outcome of a particular sliding system, the friction study could simply report "stick-slip behavior." Sometimes various sliding couples display varying degrees of stick-slip. In such instances, the friction force variability, for example, root mean square (RMS), etc., can be used as a comparison metric (Figure 11.11).

Most continuous sliding systems do not exhibit a nice plateau for friction force. There is variability with time, as shown in Figure 11.12. This is not stick-slip. It is not a harmonic vibration. It is normal random variation of friction force with increased sliding distance.

Another important fundamental of sliding friction is a concept of asperity interaction on rubbing surfaces. The friction force is really the net result of several surface interactions:

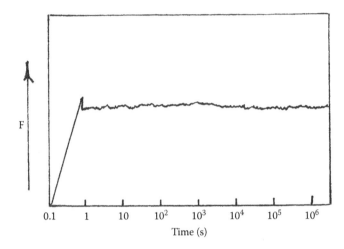

FIGURE 11.12
Normal variation of friction force over a long test time.

$$F = Fa + Ft + Ff + Fsy$$

where:

  F = friction force
  Fa = component caused by surface adhesion
  Ft = component caused by surface texture effects
  Ff = component caused by surface films
  Fsy = component caused by the sliding system conditions

There may be other components, but all continuous sliding systems have at least these components making up the friction force that is measured. The adhesion component of friction predominates when services are atomically clean and under very high loads, as in galling conditions. The role of surface texture is important when surfaces contain features that may interlock, as shown in Figure 11.13. An example would be a lathe-turned shaft reciprocating within a lathe-turned bearing bore. However, surface texture is often not the predominating component of the friction force. The early researchers blamed friction on interfering rugosities. And they may have been a significant factor at that time, since they did not have precision-rolled metals and ground ceramics, but only hand-hewn wood shapes and ceramics and their metal shapes were hammer forged or cast. Today we usually have lesser rugosities.

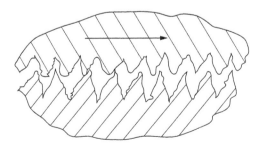

FIGURE 11.13
Retarding force (friction) caused by interpenetrating surface features.

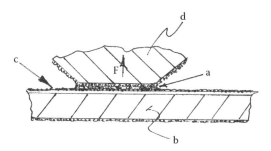

FIGURE 11.14
Stiction (F) between a magnetic read/write head (d) and magnetic media (b) due to the presence of absorbed moisture films (c) forming a meniscus (a) at the point of head/media contact.

Surface films almost always have significant effects on the friction force. Of course, any oily film lowers the friction force, as does fine separating particles that can produce rolling; they get surfaces to roll on each other. Moisture films have major influences in many tribosystems. For example, the term *stiction* was coined to describe problems that magnetic media used to have when they would set for a time in contact with a magnetic tape or disk. There would be a very high static friction coefficient, and it was also learned that it required a significant force to lift a recording head from the tape contact—it was stuck. Eventually it was learned that moisture collecting on the magnetic media contact formed a meniscus, and the surface tension of the water film was a significant component of the stiction force (Figure 11.14). This problem has all but disappeared on magnetic media because molecularly thin lubricants have been applied to prevent moisture buildup and lubricate the magnetic media.

System affects, Fsy, are probably the biggest component in any friction system. How a body slides on another body can make a very big difference because friction is a system effect. The system that controls the forces and motion is often the most important in controlling the results obtained in a material couple's frictional behavior. Figure 11.15 shows test data for the same couple's friction tested on four different test rigs. Very different

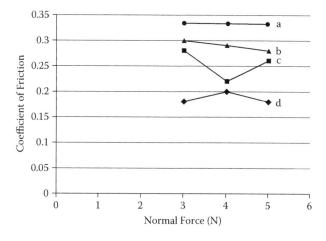

FIGURE 11.15
Variation of average coefficient of friction of the same couple (52100 steel @ 60 HRC) versus polyester (PET film) tested at three normal forces on four different test rigs: (a) = steel pin on PET disk; (b) = steel sled on PET flat; (c) = PET block on steel ring; (d) = steel capstan with PET film pulled over it (capstan test).

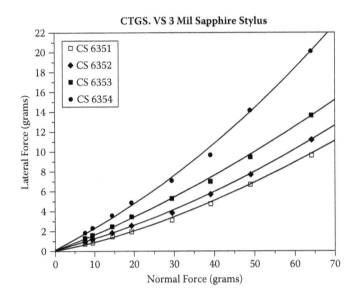

FIGURE 11.16
Friction force (lateral force) measurements made on an atomic force microscope (AFM). The coefficient of friction of a couple is often taken as the slope of the friction force–normal force curve.

answers were obtained for the friction coefficient. Thus, the system component of friction force can be an overriding factor.

Friction force measurements can often be conducted at various normal forces. There are tribology researchers who believe that the best way to measure the coefficient of friction of a mating couple is to test for friction force under varying normal forces, like was done in Figure 11.16, with the exception that the forces should go from zero to some high value. Figure 11.16 shows a typical friction force versus normal force curve produced with an atomic force microscope (AFM). The slope of the curve is considered to be the coefficient of friction for a test coating versus the stylus tip, which was a sapphire sphere in this instance. This is typical practice for nano- and microsliding of device tips against surfaces. However, in the "macro," many tribologists record friction force during galling test, where the normal force may go from 1 N to as much as 60,000 N. The coefficient of friction for countless couples in the ASTM G 98 standard galling test looks like Figure 11.17. The coefficient of friction increases with increasing normal force, and then plateaus at some low value, like 0.1. The explanation for this could be that when the contact force gets very high, the rubbing surfaces plastically deform and the resistance to motion remains about the same, while the normal force continues to increase; this makes the friction coefficient decrease. Essentially the plastic deformation of the rubbing surfaces is behaving like fluid–film separation.

A final useful friction fundamental is the use of the capstan formula. Earlier it was mentioned that there are many engineering formulas that deal with friction of ropes and other flexible substances contacting a cylinder. The traditional sled friction equations give way to the capstan equation for sliding friction, which is shown in Figure 11.18. This equation, for example, shows how much force a person will see in the bitter end of a rope if the rope is wrapped three times around a capstan and the load on the tension end of the rope is 200 pounds. It is common in line handling for boats and ships for a deckhand to stop a 40,000-ton ship with five wraps of the line around a bollard. The capstan equation is where

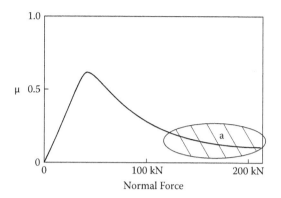

FIGURE 11.17

Typical friction coefficient behavior in metal-on-metal galling tests. Friction coefficients usually become low when galling occurs (a).

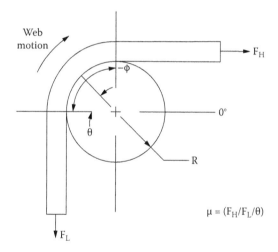

$$\mu = (F_H/F_L/\theta)$$

FIGURE 11.18

The capstan formula for the coefficient of friction of a couple that involves wrapping the flexible member around a curved or cylindrical surface.

the five wraps comes from. The more wraps, the greater the friction. The same equation applies to contact area on some types of brakes. An important concept to remember is that the net friction force developed is a function of the angle of contact in radians. This formula, along with the sled formulas, constitutes most of the mathematics normally used in coefficient of friction calculations.

Measuring Friction Force

The early friction researchers, like Leonardo da Vinci, used mostly sled techniques like the ones illustrated in Figure 11.2. Most friction measurements are made with the same

Worldwide Friction Standards (402)

Organization	
AIA	xxxxx
ASTM	xxxxxxxxxxxxxxxxxxxxxxxxxxxxxxxxxxxxxxxxx
AWS	xxxxxxxxxxx
BSI	xxxxxxxxxxxxxxxxxxxxxxxx
CED	xxxx
CEN	xxxxxxxx
CNS	xxxxxxx
CPPA	xxxxx
CSA	xxxxxxxx
CSIC	xxxxxxxxxxxxxxxx
DIN	xxxxxxxxxxxxxxxxxxxxxxxxxxxxxxxxx
DOD	xxxxxxxxxxxxxxxxx
ESDU	xxxxxxx
Ford	xxxxxxxxxxxxxxxxxxxxxxxxxxxxx
GM	xxxxxxxx
ISO	xxxxxxxxxx
JIS	xxxxxxxx
KS	xxxxxxxxxxx
NATO	xxxxxx
SABS	xxxxxxxxxxx
SAE	xxxxxxxxxxxxxxxxxxxxxxxxxxxxxx
SNZ	xxxxx
TAPPI	xxxxxxxxx
VDI	xxxxxxx

FIGURE 11.19
Acronyms for worldwide friction testing standards and the numbers of standards in each organization (circa 2000).

concept, except that the friction forces are measured by an electronic force transducer. There are many international and national standards for measuring friction. Figure 11.19 shows some of them, and they roughly fit into the categories shown in Figure 11.20. These categories mix tests with applications, but basically these data suggest the following as the most widely used tests to measure friction:

1. Sled
2. Capstan
3. Inclined plane
4. Special rigs for lubricated sliding/rolling
5. Galling test rigs
6. Scanning probe microscope (SPM)

These tests are shown schematically in Figure 11.21. The capstan test uses a force transducer to pull a web, rope, yarn, etc., over a cylinder. The inclined plane only yields the breakaway or static coefficient of friction. The lubricated test shown is a block-on-ring test

Worldwide Friction Standards

	Subject
Fabric/yarns	xxxxxxxx
Sled test	xxxxxx
Inclined plane test	xxxxxxxxx
Lubes/fluids	xxxxxxxxxxxxxxxxxxxxxxxxxxxxxxxx
Flooring	xxxxxxxxxxxx
Pavement	xxxxxxxxxxxxxxxxxx
Soil	xxxxx
Fasteners	xxxxxxxxxxxxx
Capstan test	xxxxxxxx
Brakes/clutches	xxxxxxxxxxxxxxxxxxxxxxxx
Rolling element bearings	xxxxxxxxxxxxxxxxxxx
Plain bearings	xxxxxxxxxxxxxxxx
Specific mechanisms	xxxxxxx
Magnetic media	xx

FIGURE 11.20
Categorization of tests (and relative number) listed in Figure 11.19.

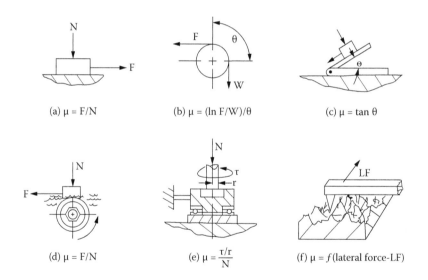

(a) $\mu = F/N$ (b) $\mu = (\ln F/W)/\theta$ (c) $\mu = \tan \theta$

(d) $\mu = F/N$ (e) $\mu = \dfrac{\tau/r}{N}$ (f) $\mu = f$ (lateral force-LF)

FIGURE 11.21
Schematics of the more commonly used friction tests: (a) = sled test, (b) = capstan test, (c) = inclined plane, (d) = block-on-ring test, (e) = galling test, (f) = scanning probe microscope with lateral force capability.

with the contacting surfaces immersed in oil. This is also a wear test, but if it is run for only a few revolutions, and there are no third bodies separating the contacting surfaces, it can yield valid friction data. There are many types of lubricated wear tests that yield friction data. The block-on-ring simulates plain bearings. There are also friction standards for lubricated rolling element bearings. These will be discussed in the next chapter. The most widely used galling test, ASTM G 98, uses a flat-ended cylinder rotating 360° on end on a fixed flat and horizontal substrate. Torque can be measured and converted to a friction

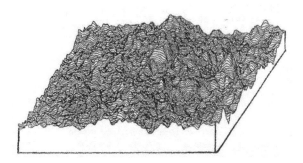

FIGURE 11.22
Surface profile scan of a 10 × 10 μm lapped surface using an atomic force microscope (AFM).

force. This test can measure the friction characteristics of mating couples under extreme contact pressures. Scanning probe microscopes are the other extreme. It can apply normal forces in the nano- and microrange as it scans across an area that may be only 10 × 10 μm in area. The scanning probe will also yield surface texture information so the lateral force (friction) can be correlated with surface features (Figure 11.22).

There are many details involved in measuring friction forces. The most used force measuring systems use wire strain gauges or piezoelectric transducers. Strain gauges have the disadvantage that significant motion is needed to produce a signal. Piezo devices have the disadvantage of electronic drift, and thus they are not suitable for continuous sliding system, but they excel in reciprocating or impact sliding systems. There are standards like ASTM G 119, that provide guidance on the ways to capture data, acquisition rates, and treatment of the data. Friction force recordings can be continuous with sampling at some designated time interval (Figure 11.23a); they can be sampled periodically during a test in process. Wear tests can be stopped when a preset friction force threshold is reached. Friction data like those shown in Figure 11.23, where (a) and (b) are typical in wear tests that yield system friction during the test. If force sampling is rapid enough, single cycles in reciprocating tests can be recorded, and the individual cycles can be plotted in 3D to produce a friction model for a test.

In summary, friction testing simply involves measuring the resistance to motion under some conditions of motion. There are many ways to record these forces and many ways to interpret the results. The correct way to test is to use a test that accurately simulates the tribosystem under study. However, always remember that a mating couple does not have a coefficient of friction. It is a system effect. A coefficient of friction measured for a couple in a particular tribosystem may not be the same as for the same couple tested in a different tribosystem.

Factors That Affect Sliding Friction

Because friction is a system effect as well as a material couple effect, the categories of effects shown in Figure 11.24 apply. Material couple intuitively affects friction. Some materials, like elastomers, are known to not slide well on most surfaces, and some polymeric materials are known to be slippery under certain circumstances, but there are many details of

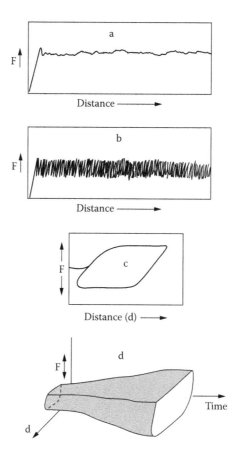

FIGURE 11.23
Typical friction force recordings: (a) = continuous sliding, (b) = reciprocating sliding, (c) = one forward and back stroke in a reciprocating test, (d) = 3D plot of reciprocating friction force loops with test time (a friction force log).

each mating material that could affect sliding friction in contact with another material. The chemical composition of both members is important, as is the processing method used to produce each member. Plastics are particularly prone to be affected by their method of manufacture. Was it molded, extruded, rolled, etc.? With metals, microstructure, heat treatment, and hardness may affect friction results. Coatings and surface treatments definitely have important friction results. However, surface condition is often thought to play a prominent role in sliding friction results. It was previously mentioned that there are many empirical and first-principle equations for coefficient of friction based upon surface texture. Before discussing these, it may be well to define what is important with respect to surface texture on any solid. Figure 11.25 shows the major features that describe a friction surface. Lay is the macroscopic orientation of surface features. A polished surface should have no lay. All surfaces produced by machining or grinding processes will have a lay. In sliding, friction lay can be important, and when laboratory tests are run, test specimens should have the same lay as the intended application.

Waviness is believed by some wear researchers to be the most important surface feature in determining the real area of contact between mating services. In the early days of building machine tool ways needed to be precise. Critical surfaces were always scraped. A thin dye layer was applied to one surface, and the two surfaces were mated to indicate

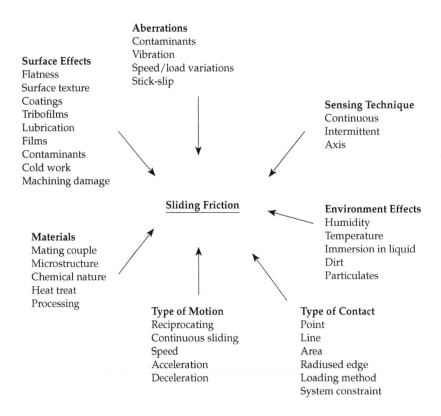

FIGURE 11.24
Some of the system effects that affect the friction characteristics of a sliding (or rolling) couple.

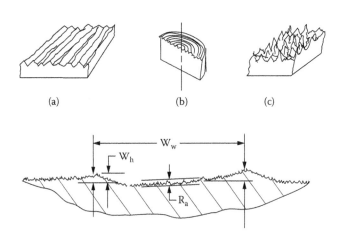

FIGURE 11.25
(Top) Major components of a surface lay: (a) = ground, (b) = lathe turned, (c) = abrasive-blasted. (Bottom) Surface texture: (R_a) = roughness, (W_w) = waviness width, (W_h) = waviness height.

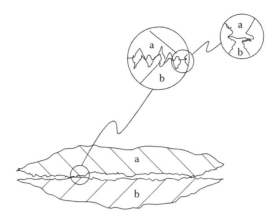

FIGURE 11.26
Plastic deformation of asperities at real areas of contact.

points of real contact. These high spots were scraped by special tools, and the process was repeated until the conflicting waveforms were minimized. The dye transfer was uniform when the scraped services were mated at the end of the process. Scraping is a process that can be used to eliminate waveforms that are undesirable on machine/fabricated surfaces. Mating contact occurs first on waveforms; however, it is commonly thought that asperities make up the real area of contact. However, they are at least partially flattened with even the lightest of loads pressing the bodies together (Figure 11.26). The problem with dealing with surface waveforms in engineering situations is that the surface texture measuring equipment used in most manufacturing entities does not yield surface waviness in its measurements because the sensors typically only record surface fluctuation over short distance, like less than 1 mm, and waveforms can have wave pitches that are centimeters or more. The waviness dilemma is not likely to be solved in the near future, so people studying friction should be aware of its importance.

Surface roughness, R_a, is the most widely used descriptor of surface texture, and it is the arithmetic average of the height of the peak to valleys on a surface over a certain distance. Root mean square (RMS) roughness is also used to describe surface roughness, and it varies numerically from R_a by about 10%. There are about 20 other mathematical relationships that are used to describe surface features, but R_a and RMS continue to be the most useful parameters. They correlate best with function. Figure 11.25 illustrates the important components of surface texture.

In lubricated systems, the separating film between rubbing surfaces is often determined by the roughness of the mating members. For this reason, many rubbing surfaces are subjected to lapping or similar processes to control lubricant film thickness, and thus wear and friction. However, in unlubricated sliding tests, surface texture may not be of overriding importance unless the mating surfaces are rough to the point where interfering rugosities of the Middle Ages are present. If there is a surface interference, the interfering features need to be sheared or deformed for motion to proceed, and this is likely to affect the friction force.

The nature of the tribosystem absolutely has a profound effect on friction results. It was shown in Figure 11.15 that the same sliding couple tested on four different test rigs produced a friction coefficient that varied from 0.22 to 0.35. This is the result of system differences—constraints, loading, motions, etc. Type of contact may also have an effect. Many laboratory test rigs use a ball as one sliding member. A ball-on-flat presents a sliding

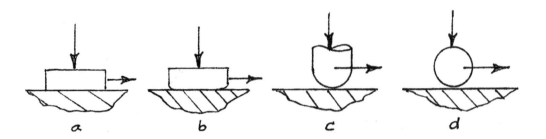

FIGURE 11.27
Some possible friction contacts: (a) = flat-on-flat, (b) = flat-on-flat with radiused edges, (c) = hemispherical rider-on-flat, (d) = cylinder-on-flat.

system that may not simulate, for example, a piston ring in a cylinder. Thus, in friction testing, a bench tester is usually advisable that simulates the type of contact encountered in the system of interest (Figure 11.27). Of course, speeds and loads should represent the application under study.

Environment is all-important. If a system under study is lubricated, friction should be measured using the same lubricant in the same lubrication conditions. Moisture from humidity in the air also almost always affects friction measurements. Sometimes moisture on surfaces can be high enough to produce meniscus effects, as we saw in stiction.

Temperature almost always has an effect on friction results. If an application is in a dryer at 300°C, the test must be run at this temperature. Metal can behave differently at temperatures above 200°C compared to room temperature, which is only 20°C or so. Plastics can behave differently with temperatures raised to only 10 or 20°C above ambient.

Films and contaminants of any type can affect friction. They need to be like those present in the application under study.

Chapter 2 mentioned friction-sensing techniques, but at this point it may be well to caution people making friction measurements not to let the computer data acquisition program dictate how friction force data are processed. As shown in Figure 11.28, a computer programmed to record over 10 sec or some other interval of time and plot average friction coefficients may miss significant differences between materials. Averaging force readings often may mask energy dissipation effects.

Finally, aberrations like vibrations and stick-slip behavior or motion variability can affect friction results. As mentioned previously, stick-slip behavior is a friction result, and it is not recommended to try to apply coefficient of friction to data obtained under stick-slip conditions. Vibration can cause contacting services to make a break contact during sliding. If the application involves vibration, do the test with that vibration. If the application does not involve vibration, then it is an aberration to suppress.

Sliding Friction Manifestations

Figure 11.29 shows the coefficient of friction (COF) of a hard steel rider (52100 @ 60 HRC) in continuous sliding on a hard steel counterface (1045 @ 55 HRC) without lubrication. The upper curve shows the variation of the COF with increasing sliding distance. The lower curve shows the COF of the same couple lubricated with light mineral oil. Note that there

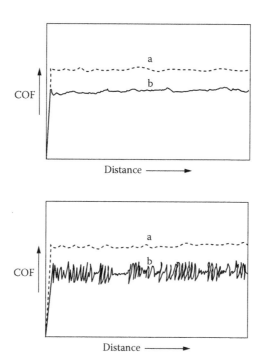

FIGURE 11.28
Couple a compared with couple b using averaged friction force data (upper). Couple a compared with couple b with unaveraged force data.

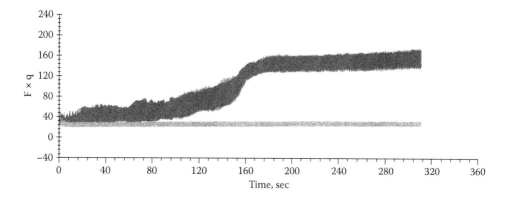

FIGURE 11.29
Friction coefficient for a hard steel rider on a hard steel counterface unlubricated (black) and lubricated with light mineral oil (gray). Pin-on-disk test rig, 2 N normal force, 0.002 m/sec, 300 sec.

is a rough incubation period in the unlubricated sliding system. Both members are adhesively interacting and a wear scar is established. Eventually the sliding contacting surfaces become separated by wear debris and the COF stabilizes. This graph also illustrates the effectiveness of lubrication in altering friction. The lubricant used in this study, light mineral oil, is not a good oil as oils go. However, it still lowered the system friction from about 0.8 to 0.1, and the incubation period lasted only seconds. Thus, lubricants can be essential to reducing sliding friction.

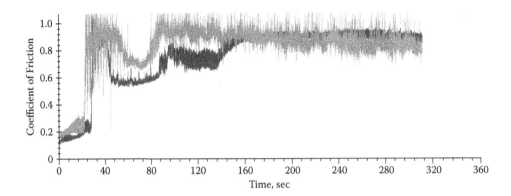

FIGURE 11.30
Friction coefficient for a hard steel rider on a soft steel counterface (black) and for a soft steel rider on a soft steel counterface (gray) using the same rig and test as in Figure 11.29. Both tests are unlubricated.

The black curve in Figure 11.30 shows a repeat of the test in Figure 11.29, only in this case the hard steel rider (52100 @ 60 HRC) was sliding unlubricated on a soft steel counterface (type 430 SS @ 95 HRB). The gray curve is the COF of soft steel versus soft steel. The hard 52100 steel rider was tempered to 95 HRB, and it slid on the same soft stainless steel used in the hard-on-soft steel test (430 SS @ 95 HRB). Note that the incubation period is rougher than for the unlubricated hard versus hard curve in Figure 11.29. These data show that there is no COF benefit for a hard–soft couple compared to a soft–soft steel couple. They both developed the same steady-state COF, which was about 0.9 compared to 0.8 for the unlubricated hard–hard sliding couple. From the practical standpoint, none of these couples (unlubricated hard-hard, hard–soft, soft–soft) are compatible; their COFs are high and system wear is significant.

Figure 11.31 shows the same test as used in Figures 11.29 and 11.30, but this time the sliding couple is hard steel (52100 @ 60 HRC) versus two different plastic counterfaces. The upper curve is for hard steel versus plasticized vinyl (PVC), and the lower curve is the COF of hard steel versus ultra-high molecular weight polyethylene (UHMWPE). These curves illustrate the effectiveness of certain plastics in reducing COF versus metals. Plasticized vinyls are typically on the high end of the COF spectrum rubbing on metals, while UHMWPE is known to be one of the slippery plastics. These test data support this contention. The steady-state COF was about 0.13, almost as good as oil lubrication. The message here is that plastics have great utility in lowering friction if the right sliding couples are employed.

Figure 11.32 shows (again) the effect of sliding velocity in lubricated sliding. These data are for a reciprocating test (ASTM G 133) of a hard–hard couple (52100 @ 60 HRC versus

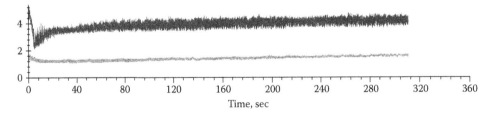

FIGURE 11.31
Friction coefficients for a hard steel rider versus two different plastic counterfaces: black = PVC, gray = UHMWPE.

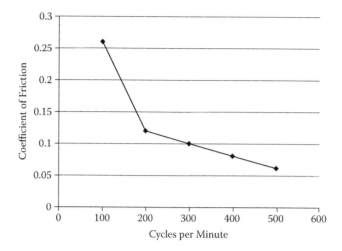

FIGURE 11.32
Effect of velocity on friction force in a reciprocating pin-on-flat test of a hard rider on a hard counterface in light mineral oil.

D2 tool steel @ 60 HRC) lubricated with light mineral oil (the same one used in the continuous sliding tests). Note that the COF decreases as the sliding speed increases. This is predicted by the Stribeck curve. The important point of this illustration is that if the goal of a laboratory test is to compare the system wear of several candidate material couples, if the test is conducted at a high sliding speed, the test surfaces do not touch, hydrodynamic lubrication exists, and the rubbing pair is continuously separated by a film of oil. In this tribosystem, a wear test should only be conducted at the 100 cpm speed where boundary lubrication exists. If you want to compare the friction and wear of sliding couples, you need to use test conditions that allow them to rub on each other.

Summary

Sliding friction is ubiquitous: it is in all machines that have rubbing contacts, it is in most of our body joints, and it is part of nature. It is the net result of changes that take place at a rubbing or rolling contact. If the contacting surfaces are getting "machined," as is often the case in solid–solid systems, the friction force and COF are measures of the energy required to produce accommodation changes on the rubbing surfaces. In the case of wearing surfaces, particles are fractured, surface texture is altered, and plastic and elastic deformations occur. In lubricated sliding systems, the machining action of friction may be reduced because the surfaces only contact at start-up and shutdown or when an operational event, such as speedup or unanticipated loading occurs. The friction force becomes the force needed to shear molecules in the lubricating film or to make atom planes slide on each other when dry film lubricants provide the separating film. The energy required to accommodate what happens when solids (or liquids) rub on each other, the co–mingling that has to happen, is friction, and design engineers must be aware of it, measure it, and deal with it. It will always be there.

Related Reading

ASTM D 1894, *Standard Test Method for Static and Kinetic Coefficients of Friction of Plastic Film and Sheeting*, West Conshohocken, PA: ASTM International.

ASTM D 4783, *Standard Test Method for Determination of the Coefficient of Friction of Lubricants Using the 4-Ball Wear Test Machine*, West Conshohocken, PA: ASTM International.

ASTM G 115, *Standard Test Method for Guide for Measuring and Reporting Friction Characteristics*, West Conshohocken, PA: ASTM International.

ASTM G 143, *Standard Test Method for Assessment of Web/Roller Friction Characteristics*, West Conshohocken, PA: ASTM International.

ASTM G 181, *Standard Test Method for Conducting Friction Tests of Piston Ring and Cylinder Liner Materials Under Lubricated Conditions*, West Conshohocken, PA: ASTM International.

ASTM G 182, *Standard Test Method for Determination of Surface Lubrication on Flexible Webs*, West Conshohocken, PA: ASTM International.

ASTM G 203, *Standard Test Method for Guide for Determining Friction Energy Dissipation in Reciprocating Tribosystems*, West Conshohocken, PA: ASTM International.

Bikerman, J.J., *Principles and Applications of Tribology*, 2nd ed., New York: John Wiley, 1961.

Czichos, H., *A Systems Approach to the Science and Technology of Friction, Lubrication and Wear*, Amsterdam: Elsevier, 2009.

Yamaguchi, Y., *Tribology of Plastic Materials*, Amsterdam: Elsevier, 1990.

Yust, C.S., Bayer, R.G., Ed., *Selection and Use of Wear Tests for Ceramics*, West Conshohocken, PA: ASTM International, 1988.

12

Rolling Friction

Without a doubt, prehistoric civilizations used rolling friction to their advantage because everyone who hunted and gathered for existence found out that is easier to roll a log than carry it, that it's easier to move heavy objects on rollers than sliding them, and that large, round stones can be rolled instead of carried. Nobody really knows when the rolling-log concept morphed into the wheel, but archaeological digs have discovered that stone bearings go back more than 10,000 years, bronze bearings go back about 3000 years, iron bearings about 2500 years, and the Romans used ball roller bearings in 280 A.D. or so. The advantage of rolling friction over sliding friction has been known for a very long time.

The importance of rolling friction can be seen by simply looking about our present-day world. Wheels are everywhere. The vehicles that clutter our roads and garages exist because of wheels. The low rolling resistance of modern-day vehicles is produced by wheels that rotate and rolling element bearings with low rolling friction, low enough to allow propulsion with the relatively small engines. A fractional horsepower engine on a moped-type vehicle can propel the vehicle to 40 mph. Thus, vehicles of all types rely on low rolling resistance from tires on pavement and roller bearings to allow the wheels to easily roll with respect to the vehicle. In addition, many internal combustion engines rely on ball or roller bearings for internal functions, such as roller cams, roller valve lifters, etc. Then there are the mass transit cars and trains that roll on tracks. Steel wheels on steel rails provide low enough rolling friction to allow high-speed trains that can hit speeds up to 200 mph (322 kmph) or more. We use the same concept of steel wheels and steel track to move heavy loads on overhead cranes in factories. Then finished goods roll down rolling element bearing conveyors to shipping docks.

Electric motors that are everywhere in our lives run on ball bearings if they operate at speeds over about 2000 rpm. Plain bearings cannot take this kind of speed. Ball and roller bearings allow the jet engines that we depend on for longer trips. Finally, gears often employ rolling between teeth. At least that is the goal. If gear teeth roll on each other at the pitch line, they will dissipate less friction and wear less than if they slide on each other. The purpose of this chapter is to present what is generally known about rolling friction and show how it is manifested in testing and in many mechanisms. The chapter objective is to give readers guidance on dealing with rolling friction from the standpoint of design engineering. This chapter will start with terms and definitions and the fundamentals of rolling friction, then discuss ways that it is measured and what these measurements look like and mean. The chapter will conclude with suggestions on dealing with rolling friction problems.

Fundamentals of Rolling Friction

Everybody thinks that they know what rolling means and a definition is not necessary. A dictionary definition of rolling is "to move forward on a surface by rotating about an

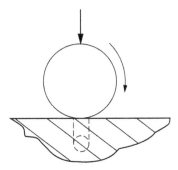

FIGURE 12.1
The dashed circle shows the no-slip part of the rolling contact.

axis." However, true rolling requires no slip in some portion of the contact area and a direction: rolling is motion in a direction of nearly spherical or cylindrical shapes characterized by no relative slip in a portion of the contact between the rolling member and the rolling counterface. The mating surfaces are at the same velocity. In cases like gears where both surfaces are in motion, the no-slip portion of the contact can be very small. A rolling sphere will have zero slip in a ring, as shown in Figure 12.1. A cylinder on a flat will have two lines of no slip. Rolling shapes other than on flat surfaces will have differently shaped no-slip regions, but there must be a no-slip region of some sort to meet our definition of rolling. There is always slip in the center of a spherical contact because of Hooke's law; for every stress there is strain. So, the center of the sphere deforms under the load that is forcing it against another surface, and so does the counterface. Deformation, even elastic deformation, requires relative motion or slip. However, at some distance from the sphere centerline, the contact stress is dissipated and no strains occur; this is the no-slip ring as shown in Figure 12.1.

Traction is a common term encountered with rolling systems. People tend to use the term synonymously with friction. There is a difference. The ASTM terms and definitions relating to wear, ASTM G 40, defines traction as "a physical process in which a tangential force (F_t) is transmitted across the interface between two bodies by dry friction or intervening fluid film resulting in motion, reduction in motion or transmission of power." Basically, friction is a retarding force and traction is a force producing or reducing rolling motion. There is also a traction coefficient that is defined in ASTM G 40 as "the dimensionless ratio of the traction force (F_t) transmitted between two bodies to the normal force (N) pressing them together."

The difference between traction and rolling friction is illustrated in Figure 12.2. If a wheeled wagon is pushed, rolling friction occurs at each wheel resisting motion. If the wheels are driven, they produce a traction force to produce motion. Mathematically, the traction coefficient is the same as the rolling coefficient of friction:

$$T = F_t/N$$

The rolling coefficient of friction is F_r/N, where F_r is the force needed to produce rolling.

It took 27 kg of force to the bumper (spring gauge pulling on the bumper) to move a 2000 kg truck at rest on smooth concrete. Thus, that tribosystem had a coefficient of rolling friction of 27/2000, or 0.0135. The traction force would be measured by force transducers on the drive wheels, and it may be different from the rolling coefficient of friction. The traction

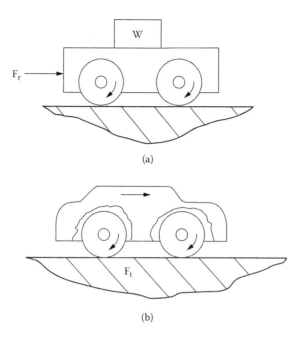

(a)

(b)

FIGURE 12.2
The difference between the rolling friction force (F_r) and traction (F_t).

coefficient may be different because the force to produce motion would be produced by the tires contacting the pavement. From the practical standpoint, *traction* and *traction coefficient* are terms used in the tire and lubrication industries, and traction and traction coefficients apply to other than rolling tribosystems. For example, automatic transmission fluids are traction fluids, and they transmit forces between sliding flat plates, not rolling elements. So, for our rolling friction discussions, we are mostly using these terms:

F_r = rolling resistance force

N = normal force

μ_r = coefficient of rolling resistance = F_r/N

What is the origin of the rolling resistance force F_r? It is generally thought that rolling resistance consists of a number of components:

Fa—adhesion

Fb—plowing

Fh—hysteresis

Ff—films (on the surfaces)

Fru—rugosities

Fe—a component due to the elastic moduli of the mating surfaces

$$F_r = Fa + Fb + Fh + Ff + Fru + Fe$$

These are illustrated in Figure 12.3.

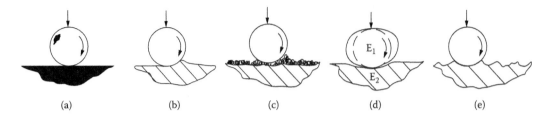

FIGURE 12.3
Components of rolling friction: (a) adhesive transfer; (b) plowing of the counterface; (c) retardation from surface films; (d) hysteresis of contacting members and elastic modulus of members effect; (e) surface rugosity effects.

Thus, the rolling resistance force is multicomponent like the sliding friction force. The material forming a rolling element can adhere to the rolling counterface or vice versa. This adhesion will retard motion. Rolling elements can plow a wave of material in their path—like casters do on carpet. This can be elastic, or plowing can mean making a furrow, like a wheel on a wheelbarrow does in soft dirt. Similarly, a sofa on casters can plow furrows in a wood floor if it is heavy enough and the casters are small in diameter. In this instance, the plowing component produces plastic deformation of the counterface. Regardless of whether these deformations are elastic or plastic and retard rolling, they take energy from the rolling member. The film component of rolling friction is the retardation caused by the rolling element having to clear away liquid or solid films to produce rolling. The force to clear away liquids or greases depends on their viscosity. This retardation component can be very significant. The lubricant in a ball bearing can be responsible for more than 50% of the friction of that bearing. If the rolling surfaces have moisture on them or are immersed in a fluid, the fluids must be displaced for rolling to occur, and this requires energy; retardation occurs. Hystereses of contacting surfaces are energy losses in elastic deflections, and they are usually minor with metal-on-metal rolling elements, but they can be a significant retardation force in plastics and rubbers. Rubbers are viscoelastic, which means that the return of strain or deflection can be time dependent. The more time dependent, the more energy required, and the more retardation to rolling. Hysteresis effects can be significant, for example, in a rolling of rubber vehicle tires on payment. The heat from the hysteresis encountered in flexing and the amount of hysteresis that occurs are functions of the type of rubber and its hardness.

The second to last component in the rolling friction force relationship, the force produced by rugosities, can be higher if you are trying to push a shopping cart on a loose stone roadway, or very minor with polished balls on finally ground raceways. The effect of rugosities is one of the reasons why ball, wheel, or cylinder diameter affects rolling friction. As shown in Figure 12.4, surface roughness may have significant effect on small diameters, but not large ones. There are models that make the rolling friction force dependent on the ball or cylinder radius. The larger the radius, the lower the rolling friction force. The rugosities that can affect rolling friction do not have to be surface texture features. They can be particles, platelets, or contaminants in bearings. Their presence will increase rolling friction force. People have tried to lubricate ball bearings with antiseizure compounds that contain soft metal particles. They mechanically plate on raceways and producer rugosities that impede rolling.

The final term in the friction force relationship, Fe, is the *elastic modulus* or hardness of the members in the rolling tribosystem. The stiffer the material couple, the lower the rolling resistance. With rubbers and elastomers, their hardness determines the rolling resistance. Solid rubber tires on an appliance truck will allow rolling a load with less effort

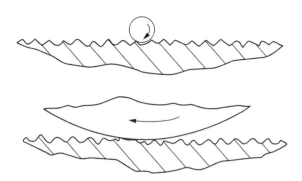

FIGURE 12.4
Effect of rolling radius on the roughness (rugosity) component of rolling friction.

than pneumatic tires. There are models in the literature that relate rolling friction directly to the elastic modulus of both members.

Thus, the rolling resistance force is dependent on at least the seven factors that we enumerated. There are more proposed by others, but the ones that we listed above are commonsense factors. They can be observed in action on many of the wheeled devices that people use daily. The little wheels on suitcases would go better if they were bigger. Also, we know that narrow bicycle tires roll easier than fat knobby ones (Figure 12.5). This is the rugosity effect. Pulling a wheeled suitcase on a dirt path takes more effort than on a smooth airport aisle. A billiard ball rolls better than a foam rubber ball of the same size; this is the elasticity/hardness factor. The relative importance of these different forces depends on the tribosystem involved because all forms of friction are system dependent.

What else is known about the nature of rolling friction? The rolling friction force is proportional to the mass and downward force on the rolling object, but as in sliding friction, there may be an exponent on the load term: F_r is proportional to (load) raised to the 1.2 to 2.4 power.

Also, like sliding friction, the starting friction μ_{rs} is higher than the rolling kinetic friction coefficient μ_{rk}, and the kinetic rolling friction coefficient may be velocity dependent. Rolling on wet or lubricated surfaces can be easier than on dry unlubricated surfaces if the

FIGURE 12.5 (See color insert.)
Bicycle tire choices. Which has lower friction with hard roadways? Which one in dirt?

lubricant reduces the rolling friction component produced by slip outside the zero-velocity no-slip area of the contact.

Rolling friction directly depends on the radius of the rolling element:

$$F_r \text{ is proportional to } 1/R$$

where R is the radius of the rolling element.

Laboratory experiments have shown that plastic rolling elements do not roll on steel or plastic races as well as metals roll on themselves. All-plastic rolling element bearings are available for use in corrosive environments, but they do not spend freely like their metal counterparts. This is probably the result of the elastic modulus/hardness component of the rolling friction force.

Similarly, laboratory experiments have shown that rolling experiments with steel balls on an inclined plane made from steel with different surface finishes demonstrate the rugosity component of rolling friction force. The roughness effect did not persist with almost polished surfaces (less than 0.1 μm Ra), but the effect persisted to roughness of several micrometers. Rolling element bearings like ball bearings comprise one of the most significant implications of rolling friction. Researchers in this area have created models that take into account about a half dozen factors that affect the friction of a finished bearing. Friction will be high at low speeds where the roles of seals and sliding are most noticeable. When separating oil films are established, the viscosity of the lubricant is likely to have a significant effect. Friction reduces when the grease does its job and separates the rolling elements. When speeds increase further, the rolling friction increases due to breakdown of the favorable separating film thickness. On the nanoscale, most friction testing is performed with scanning probe microscopes, and they use sharp styli, usually smaller than 50 μm in diameter, to sense friction force. They do not use rolling elements.

In summary, the rolling friction force is used just like the sliding friction force to calculate rolling and coefficient of friction:

$$\mu_r = F/N$$

There is a static and a kinetic coefficient, just as the sliding and the rolling friction force have components due to the adhesive tendencies of the two materials, the plowing tendency of the rolling element on the counterface, the influence of surface films, hysteresis, surface rugosities, and the hardness/stiffness of the couple. Of these, the stiffness and rugosity components are almost always significant, whereas the others may or may not be significant, depending on the rolling tribosystem.

Testing for Rolling Friction Characteristics

There are not as many standard tests for rolling friction as there are for sliding friction. In fact, the lack of test methods for the friction of ball bearings led to the development of ASTM G 182, which is an inclined plane test (Figure 12.6) to measure the breakaway friction of rolling element bearings—to essentially measure the free spinning of various ball bearings when used for pivots and the like. There is sliding friction from seals and shields that typically affect rolling friction on ball and rolling element bearings that are used at

FIGURE 12.6
ASTM G 182 inclined plane test for starting friction of rolling element bearings.

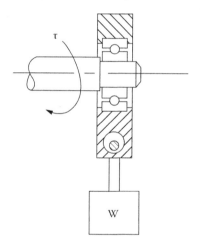

FIGURE 12.7
Schematic of a ball bearing tester that measures torque (τ) at different speeds, loads (W), and revolutions.

low speed, intermittent, oscillating, or partial revolution. Test rigs are commercially available that measure torque to turn the inner race of a rolling element bearing while the other race is fixed (Figure 12.7). Various speeds and loads can be applied to the bearing to produce a series of curves that characterize the bearing friction characteristics.

Tires rubbing on pavement are simulated by trailer-type test rigs that are drawn by vehicles on typical interstate and other types of roadways. Some rigs have water spray capability so that friction drag can be measured wet and dry. One way that friction coefficients are measured is to put a known weight on the towed vehicle, break wet or dry, and measure the force increase on the vehicle tow bar. Laboratory dynamometers are used to measure friction of tires. However, on a dynamometer, the tires rub on rollers, not actual pavement. This can be a problem. The U.S. Society of Automobile Engineers publishes a standard test to measure the rolling friction of a complete vehicle, but this also was done on a dynamometer. The vehicle is brought to various rolling speeds, for example, 50 mph, 60 mph, etc., and it is then allowed to coast from this speed. The coasting response is converted to a rolling friction rating.

There is an ASTM standard for rolling characteristics of balls rolling on a flat horizontal surface. The test can apply to ball bearings, golf balls, tennis balls, billiard balls, and even bowling balls. In this test, a ball is fixed at the top of an inclined plane of a certain height. It is allowed to roll down the plane, and the distance rolled after exiting the plane is measured. The ratio of the height of the ball above the horizontal plane to the distance rolled on the plane is the coefficient of rolling resistance (Figure 12.8). Figure 12.9 shows rolling resistance data on golf balls. The same concept has been applied to measuring the rolling resistance of vehicles, any type. The vehicles simply roll down a standard incline (hill) of known height above the horizontal roadway and the same ratio is used (Figure 12.10).

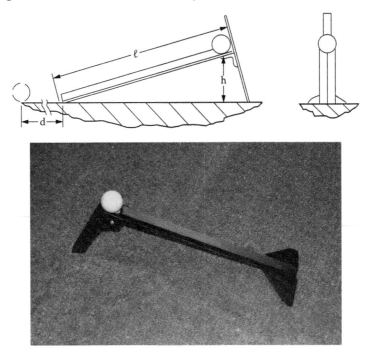

FIGURE 12.8
Schematic of ASTM G 194 test for rolling resistance of various surfaces: coefficient of rolling resistance, CORR = h/d.

Surface	Coefficient of Rolling Resistance	Rolling Resistance Number
2 in. high rough	0.6	60
Wet level green	0.2	20
Same green dry	0.109	10
Smooth concrete	0.025	3
Hardwood flooring	0.019	2
Low-pile carpet	0.1	10
Stone (pea) walkway	0.25	25
Macadam walkway	0.083	8

FIGURE 12.9
Rolling resistance test data for the same golf ball rolling on different surfaces (ASTM G 194). The higher the rolling resistance number, the higher the resistance to rolling.

FIGURE 12.10
ASTM G 194 rolling resistance test applied to vehicles—rolling down a "fixed" hill in neutral with the engine running. Test vehicles and the standard hill are shown.

The vehicle has the engine running and the transmission is in neutral; no brakes are used. Figure 12.11 shows rankings from this test.

ASTM G 115 displays scores of standard tests that measure friction of various mating couples in some fashion. Most are sliding tests, but as has been pointed out in this chapter, there is no difference between the rolling coefficient of friction equation and the sliding coefficient of friction equation. The traction coefficient is calculated like the rolling coefficient of friction, but the concept is different in that direction is a force vector at the rolling contact with the rolling surface.

One of the most popular tests for studying traction fluids and lubricants in general is the rotating sphere tangentially loaded on a rotating disk (Figure 12.12). The test couple is a disk in contact with a rotating sphere. Each member's speed is controlled independently, so the couple could have no slip or 50% slip, or whatever number the tester desires. The materials can be whatever the user wants. This test measures the traction force between the ball and the disk in any slip ratio, and the traction coefficient of a fluid can be measured. Users can also study and measure the separating film thicknesses.

Finally, very large dynamometers can be used to measure the traction characteristics of train wheels versus tracks. Sometimes just pulling loaded cars on tracks is used to

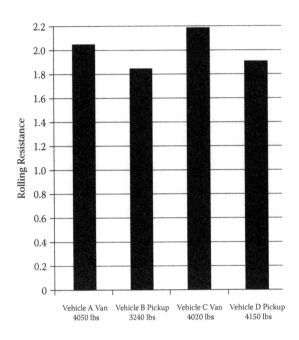

FIGURE 12.11
Rolling resistance of four random vehicles using a modification of ASTM G 194.

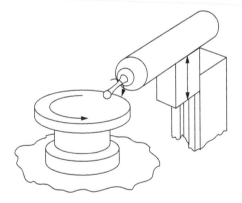

FIGURE 12.12
Test device that can produce various amounts of departure from pure rolling (velocity of ball = velocity of disk) to 20%, 50%, etc., slip at different speeds and loads. Lubricant film thickness can be measured on such a rig.

measure the rolling friction of the cars. Friction between engine wheels and tracks is a concern in determining the number of cars that can be pulled, as well as effects of grades on wheel slippage.

Overall, rolling friction testing is quite a specialty, and most testing is performed in laboratories of companies that make the rolling equipment. The ball and roller bearing people have tests that they trust, as do the train and crane rail manufacturers and the sports equipment manufacturers, but the tests that we described present the flavor of rolling friction testing. The most used tests are illustrated in Figure 11.21.

Rolling Friction Manifestations

One of the most important parts of any professional golf tournament is green speed, the resistance (or lack of) of golf greens to rolling of the golf ball on the putting surface. Green speed is controlled by the length of blades of grass on the green, the type of grass, and how the grass is mowed. For tournament play, the grass is cut to a designated speed using the rolling friction gauge called the Stimpmeter. The ASTM G 194 inclined plane test discussed in the previous section is analogous to the Stimpmeter. It is just a much larger inclined plane, and it is used to measure the distance that a golf ball travels in feet after leaving the end of the Stimpmeter ramp. For normal play, the greens may be cut to a Stimpmeter value of 8 (8 ft traveled after leaving the ramp). For tournament play, the green speed may be increased to 13 or 14 (14 ft traveled after leaving the Stimpmeter ramp). And this is documented as the speed of the putting green at that time, at that place, and under those environmental conditions. A high green speed makes putting incredibly difficult. For a downhill putt, for example, the putter of strike must be as weak as a tap to a baby's cheek. The gauge used in the ASTM G 194 test can be used on golf greens to measure rolling resistance of a golf ball (Figure 12.8). The rolling resistance number is simply the coefficient of rolling resistance multiplied by 100. Data obtained with this gage show that rolling on a dry green may have the rolling resistance number 10; wet on the same green, the rolling resistance number may be 20. It may be 60 for 2 in. high grass. If this number were applied to a green cut to a Stimpmeter reading of 13, the rolling resistance number might be 3. Thus, a simple ramp device can yield valuable information for rolling of sports balls or, in industry, it can be used for anything that rolls, like apples, oranges, etc. They can be classified with the ramp type gauge. Figure 12.11 shows rolling friction data on golf balls on different surfaces. Figure 12.8 presents coefficient of rolling resistance data on four test vehicles, the ones that are shown in Figure 12.9. Theoretically, the heaviest vehicle should have rolled the farthest because it had the highest potential energy (mass × height). However, this test takes into account all friction sources on the vehicle, the tires versus pavement, the bearings, the seals, all of the parts that rotate during rolling. The wind was calm during testing, so aerodynamic drag was not a factor. Wind can contribute at higher speed, but its contribution at speeds up to 10 mph is negligible. The winner of the tests in the illustration was the lightest vehicle. These kinds of tests can produce relative rolling resistance data that may help solve some tribological problem.

In a related study, a bicycle was tested with various inflation pressures in its tires. The data, which are shown in Figure 12.13, suggested at least a 10% improvement in rolling friction resistance to be achieved by a 2 Pa increase in tire pressure. Truck studies have shown a 10% improvement in mileage by increasing tire pressures 5 psi over the manufacturer's soft-ride recommendation. The higher pressure was still 5 psi below the tire manufacturer's recommended maximum inflation pressure. Thus, the world's usage of gasoline and diesel fuel could be reduced by 10% by simply lowering the rolling resistance of tires. This can be done with no capital costs and no taxes—only lifestyle changes, just maintaining a certain pressure in vehicle tires.

Figure 12.14 shows typical rolling friction data (from torque) on a rolling element bearing using the test shown in Figure 12.15. The friction at low rpm shows the effect of rubbing between balls and separators and shields. It also was affected by oil viscosity; this reduces as speed increases due to heating from the rubbing friction. Data like these are

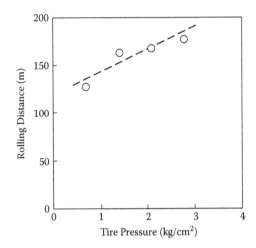

FIGURE 12.13
Effect of tire pressure on the rolling friction of a bicycle. The distance rolled from the top of a standard starting ramp (3.7 m high), same bicycle, same rider, same conditions.

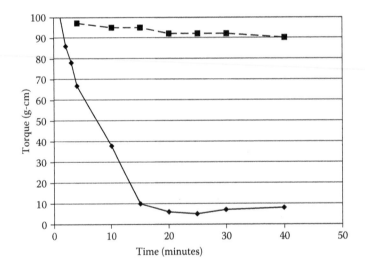

FIGURE 12.14
Typical rolling friction data (torque) from a rolling element bearing tester. The dashed line is for an unlubricated bearing; the solid is for a greased and sealed bearing.

used to profile the rolling friction of ball bearings. Many times acoustic emission is used instead of force as the test metric. When rubbing and other aberrations become significant, noises are produced that can easily be sensed by sensitive microphones.

Figures 12.16 and 12.17 demonstrate two other rolling friction concepts:

1. The effect of diameter of the rolling element
2. The effect of counterface roughness and flatness

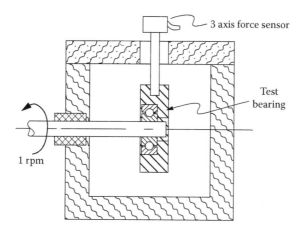

FIGURE 12.15
Test rig for measuring bearing friction at low radial forces and low speeds (for pivot applications).

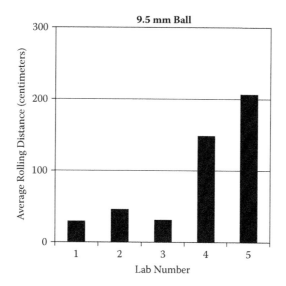

FIGURE 12.16
Rolling distance of a 9.5 mm diameter steel ball launched from a standard ramp on granite inspection tables.

There were interesting results from interlaboratory tests used in studies related to the ASTM G 194 inclined plane test. Five laboratories tested rolling of two different sized steel balls down the same ramp, using the same balls, and the distance rolled was much greater with the larger balls. Also, these labs used a granite inspection table as the rolling surface, but all of these inspection tables were made by different companies, and some were smoother than others. The travel distance was about the same in the labs using the smooth granite inspection tables, but was lower for those with nonpolished tables, thus demonstrating the effect of surface roughness or rugosities.

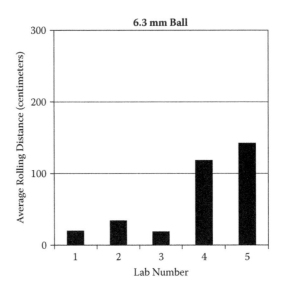

FIGURE 12.17
Rolling distance of a 6.3 mm diameter steel ball launched from a standard ramp on granite inspection tables.

Dealing with Rolling Friction

Figure 11.1 summarizes the types of friction that had been discussed in Chapter 11 on sliding friction, and Figure 12.18 summarizes what was presented in this chapter on rolling friction. Sliding friction is shown to be affected by the nature of the rubbing surface, third bodies, and separating fluids/films. These factors also affect rolling friction. If the goal is to reduce system rolling friction, there are many options. The seven factors shown to be components in the friction force equation can be altered.

A very important aspect in dealing with friction, both sliding and rolling, is to continuously monitor forces and analyze the tribosystem. Figure 12.19 shows friction force recordings for reciprocating ball-on-flat friction tests. The goal was to compare the friction characteristics of three different polymeric coatings, a, b, and c, versus hard steel under reciprocating conditions. The test was for 10,000 cycles, 10 N normal force, 52100 steel at 60 HRC rider, 10 mm stroke, 5 Hz. If the average friction force were programmed as the output of the data acquisition system, the three coatings would have shown the same average coefficient of friction. However, all three coatings wore under the test conditions. Coating a wore less than coatings b and c, and coating c wore through early in the test. If this specimen were allowed to continue with the test, the friction data would be mostly for hard steel on the coating substrate. Figure 12.19 illustrates why it is preferred to continuously record and analyze forces rather than let the computer average force data. In the case of reciprocating motion, the area of the friction force traces as shown in Figure 12.19 can serve as a measure of the frictional energy dissipated by each coating tribosystem. Friction energy losses can be quantified. The ASTM G 203 standard guide shows how to do this.

To conclude this chapter and the discussion on sliding friction, we list some rules of thumb that may be helpful in dealing with sliding and rolling friction. However, the most important message is that friction in tribosystems should not be ignored or taken lightly. It is often more important than wear and erosion from the economic standpoint.

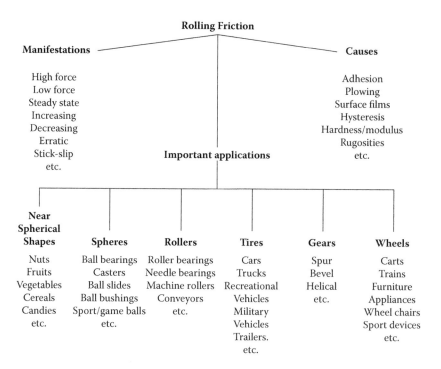

FIGURE 12.18
Applications where rolling friction may be an important, even limiting factor.

Friction Rules of Thumb

- Lubricated sliding systems usually produce friction coefficients near 0.1 under boundary lubrication conditions, and lower than this under hydrodynamic or elastohydrodynamic conditions.
- Rolling element bearings typically produce friction coefficients of less than 0.05.
- Air bearings may produce friction coefficients of 0.01 or less.
- Unlubricated sliding of clean metals usually produces friction coefficients in the range of 0.4 to 0.8.
- Plastic-to-metal couples usually have friction coefficients in the range of 0.2 to 0.4.
- Metal-to-rubber couples typically produce friction coefficients in the range of 0.8 to 3 or more.
- Metal-to-metal friction under pressures approaching their yield strength produce friction coefficients of about 0.1.

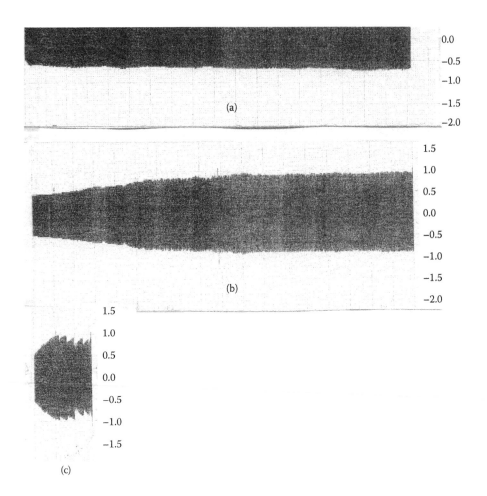

FIGURE 12.19
Friction force recordings on three different coatings versus hard steel in reciprocating sliding. The force recordings start at the left. Coating (c) failed early in the 10,000-cycle test.

Related Reading

ASTM G 143, *Standard Test Method for Measuring Rolling Friction Characteristics of a Spherical Shape on a Flat Horizontal Plane*, West Conshohocken, PA: ASTM International.

Blau, P.J., *Friction Science and Technology*, New York: Marcel Dekker, 1996.

Bowden, E.P., Tabor, D., *Friction and Lubrication of Solids*, Clarendon Press, 1950.

Buckley, D.H., *Surface Effects in Adhesion Friction, Wear and Lubrication*, Amsterdam: Elsevier, 1981.

Buckley, D.H., Dugan, M., *Measurement of Rolling Friction*, Cleveland: NASA, 1983.

Budinski, K.G., *Guide to Friction, Wear, and Erosion Testing*, MNL56, West Conshohocken, PA: ASTM International, 2007.

Pusson, B.M., Tosatto, J., *Physics of Sliding Friction*, Dordrecht: Elsevier, 1996.

Rabinowicz, E., *Friction and Wear of Materials*, 2nd ed., New York, 1995.

Singer, I.L., Pollack, H.M., Eds., *Fundamentals of Friction: Macroscopic and Microscopic*, Dordrecht: Elsevier, 1992.

13

Materials for Friction, Wear, and Erosion

This book emphasizes that there are various modes of wear, erosion, and friction, and it should be clear at this point that dealing with abrasion problems requires a very different solution than dealing with cavitation problems. However, there are rather limited options that engineers and maintenance personnel have in fixing an existing tribology problem or a perceived one. They can try to identify a material that resists a particular mode of wear or erosion without a coating or surface treatment, or they can try to identify a substrate and coating or treatment that will do the job. This chapter is about the former: materials that can solve tribology problems unaided; they do not need to be coated or surface treated, but may require ordinary heat treatments for through hardening. They are materials that can be purchased "off the shelf" and used.

Figure 13.1 lists some materials that could be used without the help of surface alterations. This list gets shorter each year because there are fewer manufacturers due to regulatory plant closures and consolidations produced by the "global economy." Materials for tools and machines basically fall into the categories shown in Figure 13.1. Sometimes these materials can be used as received; some may require a heat treatment to make them suitable for a particular mode of wear or erosion. Cost often necessitates making parts from the cheapest material that will be suitable to accept a surface engineering process to enhance the wear/erosion resistance of the surface-like carbon steel covered with a hard-facing. That kind of situation is covered in the next chapter. The purpose of this chapter is to list materials that have successfully been used in tribology applications in the past.

Ferrous Metal Alloys

There are thousands of ferrous (based on iron) metal alloys that can be used for wear and erosion applications, but the most used are shown in Figure 13.1. This illustration also lists some specific ferrous alloys that have utility in wear and erosion applications. Many machines and devices use low-carbon steels in as-received condition for tribocomponents simply because they are the lowest-cost material. Skids on household snowblowers are usually made from low-carbon steel, and they are exposed to severe abrasion from walkways, but usually last the life of the unit, which is only several seasons. Household door hinges are similarly made from low-carbon steels, and they must last for decades. They wear, but the rate is low and homeowners do not complain because they still work in the worn condition. These parts last as long as intended by their designers.

Type 1080 steel, on the other hand, as a completely pearlitic structure (Figure 13.2), is found to be quite suitable for low-stress abrasion wear in tillage tools without case hardening. This same material is used in springs and other devices that are subjected to metal-to-metal wear. The alloy steels listed in Figure 13.1 are mostly used for structural applications and the like, but many people use them for gears and similar uses that are

Engineering Materials for Wear and Erosion Applications

Metals	Plastics	Ceramics	Composites
Carbon steels	Phenolics	Silicon carbide	Reinforced PF
1040	Polyamides	Aluminum oxides	FRPs
1080	Polyamide-imide	Zirconias	CF-reinforced epoxies
Tool steels	Fluorocarbons	Silicon nitrides	Cermets/carbides
D2	Acetals	Etc.	Etc.
S7	Polyimides		
M2	PPS		
A11	UHMWPE		
Cast irons	PEEKs		
Gray	Etc.		
Ductile			
White			
Copper alloys	**Elastomers**		
Bronzes	Urethanes		
Be coppers	Silicones		
Nickel alloys	Polychloroprenes		
Monels	SBRs		
NiCrB alloys	Etc.		
Cobalt alloys			
Stellites			
Alloy steels			
4140			
4340			
8620			
Nitralloys			
Stainless steels			
420+V			
440C			
Etc.			

FIGURE 13.1
Engineering materials that can be used for erosion and wear.

exposed to wear situations. The low-carbon alloy steels 9310 and 8620 are commonly case hardened (usually carburized), and in that condition have utility in lubricated sliding wear applications. Type 52100 steel is the standard steel ball and roller rolling element bearings; it is usually hardened to 60 HRC.

The special steels listed each have different applications, but nitriding steels are particularly useful, and they are designed to accept nitriding to produce hardness that can be 70 HRC. The maraging steels contain 18% or so nickel, and they have high strength but little wear resistance. They need surface engineering as weathering steels (they form adherent rust outdoors), prehardened steel (30 HRC), and high-strength low-alloy (HSLA) steels. Manganese steels contain about 13% manganese in their composition, and this makes them rapidly work under impact wear applications (Figure 5.15). They are popular for use in track/rolling wear and impact wear applications like railroad frogs.

Stainless steels are often used in tribocorrosion applications, and the austenitic 300 series stainless steels are used in pumps and the like where they are subject to slurry or liquid

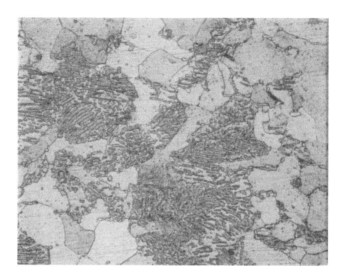

FIGURE 13.2
Pearlite in a matrix of ferrite grains. Pearlite is a lamellar structure of harder iron carbide in ferrite that provides wear and erosion better than ferrite.

impingement erosion. However, they usually need surface engineering assistance. Type 440C stainless steel can be hardened to 58–60 HRC, and it is used for balls and rollers in rolling element bearings that are used in corrosive situations. Type 420 stainless steel can be hardened to 52–55 HRC, and it is used for knives and the like. Most household silverware knives are made from it, as are most cavities for plastic injection molding. Type 17-4 PH precipitation-hardening stainless steel can be age–hardened to 43 HRC after a simple heat treatment, but it needs surface engineering for most wear applications.

Cast irons are widely used for internal combustion engine cylinders and all sorts of devices that see lubricated sliding wear. Ductile irons, like 60-45-12, are used for lubricated wear applications. White irons can be cast at 60 HRC, and they are widely used for abrasive wear and tribocorrosion applications.

Finally, tool steels are the gold standard for materials for tools that form and shape other materials: punches, dies, saw blades, milling cutters, shear blades, etc. They can be abrasion resistant and have reasonable metal-to-metal wear resistance. The feature that sets this family of steels apart from alloy steels and carbon steels is that some contain significant percentages of alloy carbides in their microstructure. Tools steels usually contain 5% or more alloy elements like chromium, molybdenum, nickel, etc., and these alloy elements tend to combine with the relatively high carbon content (about 1%) to form chromium carbides, molybdenum carbides, tungsten carbides, etc. Tool steels are available with a wide variety of microstructures (Figure 13.3). These carbide structures are often engineered for particular wear applications. For example, grades, such as A7, contain massive carbides for abrasion resistance. Type A11 contains fine carbides to hold cutting edges. The M and T series are called high-speed steels because they allow faster cutting speeds when they are used for machining other materials. Also, these are the hardest of all tools steels when properly hardened. They can have hardnesses of 65 to 68 HRC. The other tool steels in general have a maximum bulk hardness of 62 HRC.

Some tools steels contain low-carbon and low-alloy content and few carbides (P series); some contain medium carbon and no carbides, but alloy elements that allow their use at elevated temperatures (H series). Both are not well suited to abrasive wear applications

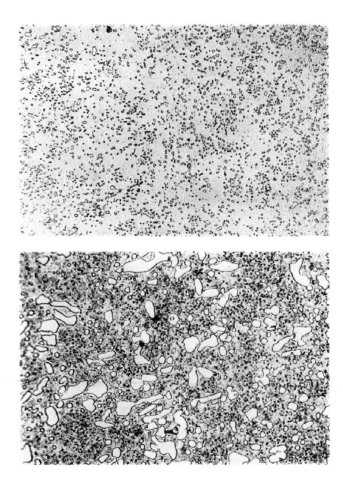

FIGURE 13.3
Microstructure of two tool steels showing the range of hard phase (carbides) from fine in type 01 to massive in type D3 (both photos have a magnification of 750×).

because of their low carbon content. Some tool steels are intended for impact applications (S series). Type S7 is widely used to resist impact wear (at 57 HRC). However, it often needs surface engineering assistance.

In general, tool steels are very useful in solving wear and erosion problems without surface engineering. The designations of the tool steel grades used in the United States are shown in Table 13.1. Usually the steels are used hardened to their normal working hardness, and this varies by grade. The grades listed in Figure 13.1 have significant utility in wear and erosion applications. The M50 grade is used for rolling element bearings, type S7 is used for impact wear, and types D2 and A11 have countless uses in tools that work other materials.

Nonferrous Metals

Aluminum alloys are extremely popular for machine components because they are easy to machine (drill, turn, mill, saw, etc.). The mechanical properties that make them easy to

TABLE 13.1

Designation Symbols for Tool Steels Used in the USA

Class	Symbol	Discriminating Factor
Water hardening tool steels	W	Carbon is the major alloying element
Cold work tool steels	O	Oil hardening
	A	Medium alloy air hardening
	D	High carbon, high chromium
Shock resisting tool	S	Medium carbon, high toughness
Mold steels	P	Low carbon, good fabricability
Hot work tool steels	H1–H19	Chromium types
	H20–H39	Tungsten types
	H40–H59	Molybdenum types
High speed tool	M	Molybdenum types
Steels	T	Tungsten types
Special-purpose tool steels	L	Low-alloy types

Note: There are a number of specific steels under each letter. Each has a different chemical composition and property.

machine (low modulus, low hardness, low to medium strength, etc.) make them also prone to abrasive and metal-to-metal wear. There are not many places where they work as tribocomponents in as-received form. However, when aluminum alloys are hardcoated, they can have wear characteristics that rival many other materials for use in tribosystems. Hardcoating and other aluminum treatments will be discussed in Chapter 14.

Copper alloys, the bronzes in particular, are widely used in tribocorrosion applications for pumps and valving. Grades can be matched to resist sliding wear. Some bronzes were invented to specifically resist cavitation erosion when used in ship propellers. Powdered metal (P/M) bronzes may be the most widely used plain bearing material on our planet. They are used on most fractional horsepower motors, and these motors are everywhere. Ideally, the shafts that rotate in P/M bearings will be hard steel and the bronze is oil impregnated. P/M bronzes can also be used for many tribocomponents other than bushings and plain bearings.

Nickel is a soft galling-prone metal, but when alloyed with other metals, it can be used for tribocomponents. Alloyed with copper, it forms monels and copper-nickel-zinc alloys, which can have utility and tribocorrosion applications. Many superalloys for jet engines and the like are nickel based, and coatings are usually needed on rubbing services. However, when nickel is alloyed with boron and chromium, it forms a family of hardfacing alloys that have great utility in galling and abrasion resistance. They will be discussed in Chapter 14.

Cobalt is a soft, malleable metal like nickel. In fact, the two often come from the same ore deposit. Cobalt has a lower melting point than nickel, and it is often used for welding hardfacing (Chapter 14). However, for almost 100 years, a family of cobalt alloys called Stellites have been widely used in all sorts of wear applications, and some are available in wrought form and most are available as castings. Their hardnesses range from about 40 to 60 HRC. They are corrosion resistant, and thus widely used for tribocorrosion applications. Some stellite grades are widely used for galling applications.

Titanium alloys are notorious for having poor wear characteristics—most modes. However, because of its unusual strength, stiffness, light weight, and corrosion resistance, titanium is widely used for body implants where it may be the counterface for a shoulder, knee, or hip. These prosthetic devices are usually made from 6% aluminum, 4% vanadium alloy, and age hardened to 36 HRC. As the counterface in an artificial joint, it usually rubs against ultra-high molecular weight polyethylene (UHMWPE). Coatings and treatments can be used to enhance their wear resistance versus UHMWPE. An unusual feature of titanium (6Al 4V) is in a galling situation; it does not form excrescences as do many metals under galling conditions. However, it often adhesively transfers to other counterfaces. This may or may not be as undesirable as excrescence formation, depending on the application. The pure grades of titanium gall in the normal way with excrescence formation. Untreated, uncoated titanium alloys all have poor abrasion resistance, even at 36 HRC.

Zinc has been used for tribocomponents for decades in the form of die-cast gears, knobs, pawls, etc. It is great for cast-to-shape parts. However, even lubricated, zinc alloys are not as long lasting as tribocomponents. Automobile carburetors were made from zinc die casting for decades, and all carburetors had many moving parts, but most required surface engineering to work properly. Zinc is soft; it has a pseudoelastic modulus and it is not at all abrasion resistant. Thus, its use in tribocomponents is limited.

Babbitts, alloys of tin, and other lead metals are cast soft alloys that are widely used for plain bearings for large machines like steam turbines. They require hydrodynamic lubrication to survive. If used unlubricated, they may only last minutes. These soft alloys are used for these critical applications because their softness allows conformability with the rotating shaft and embeddability if debris or contaminants get it into the sliding system. They get embedded rather than cause seizure by reducing the bearing's running clearance.

Ceramics/Cermets

Ceramics are inorganic compounds with crystalline structures. There are clays that can be used to form rigid solids, but they are not engineering ceramics. The ceramics that have utility as solids to be used in tribosystems are:

- Aluminum oxide
- Silicon carbide
- Silicon nitride
- Zirconium oxide

These ceramics have hardnesses greater than metals, and the Rockwell hardness scales do not apply to them. The SI hardness unit for these materials is gigapascals (GPa), as shown in Figure 13.4. The hardest of the group is silicon carbide. It is very brittle and friable, so its use is limited to applications where these limitations are not a problem. Aluminum oxide is used to rub against ultra-high molecular weight polyethylene in human joint implants, and it is used as wear tiles for abrasion and many other tribocomponents. It cannot be machined, only diamond ground; thus, this limits its applicability. Silicon nitride is becoming popular as rolling elements in rolling element bearings. It is claimed by some bearing manufacturers to be superior to 52100 steel in bearing fatigue life. Zirconia can be

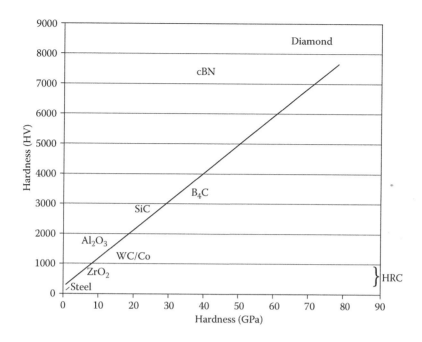

FIGURE 13.4
Conversion of Vickers hardness (HV) to hardness expressed as a pressure/stress in gigapascals (GPa).

easily molded into part shapes and is used for many different wear parts. However, it is not much harder than fully hardened tool steels, so its use is usually limited to tribocorrosion applications where its corrosion resistance is a desirable property.

Cermets by definition are ceramics with a metal binder (*Cer* for "ceramics," *met* for "metal"). The most widely used cermet is cemented carbide. However, the U.S. manufacturers of these materials tend not to call them cermets, even though they are. They use the *cermet* term for materials with high metal binder contents, like 50% and above.

Cemented carbides are carbide-metal composites, mostly of tungsten carbide particles, bonded together with a metal binder, usually cobalt with various carbide sizes from nanosize to microsize to millimeter size. Binder percentages are usually in the range of 6 to 16%, but they can be lower, and there are even binderless grades. Figure 13.5 shows typical cemented carbide grades and their U.S. designations. The typical designation system is to list the chemical composition of the carbide phase (like WC for tungsten carbide), followed by weight percent binder, followed by the chemical symbol for the binder (like Co for cobalt): WC/6Co. Some European and other countries call these materials hard metals. Cermets tend to have tool steels and stainless steels as the binder metal, and the carbide phase is often titanium carbide. The typical microstructures of cemented carbides are shown in Figure 13.6. A typical cermet microstructure is shown in Figure 13.7.

Cemented carbides are extremely useful for sliding wear and abrasion applications. They have hardnesses above 20 GPA and very high compressive strengths (greater than 400 MPa). The different grades are often researched for wear applications. Unfortunately, the cobalt binder in cemented carbide is not very corrosion resistant, and this hampers applications for tribocorrosion problems.

An Example of a Carbide Grading System

Use Category	Code	Recommended Application	Composition[a] WC	TiC	TaC	Co	Hardness[a] (RA)	Transverse Rupture Strength[a] (MPa)
Machining of:	C-1	Roughing	94	—	—	6	91	2000
Cast iron	C-2	General purpose	92	—	2	6	92	1550
Nonferrous	C-3	Finishing	92	—	4	4	92	1520
Nonmetallic material	C-4	Precision finishing	96	—	—	4	93	1400
Machining of:								
Carbon	C-5	Roughing	75	8	7	10	91	1870
Alloy	C-6	General purpose	79	8	4	9	92	1650
Tool steels	C-7	Finishing	70	12	12	6	92	1750
	C-8	Precision finishing	77	15	3	5	93	1180
Wear applications	C-9	No shock	94	—	—	6	92	1520
	C-10	Light shock	92	—	—	8	91	2000
	C-11	Heavy shock	85	—	—	15	89	2200
Impact applications	C-12	Light	88	—	—	12	88	2500
	C-13	Medium	80	—	—	20	86	2600
	C-14	Heavy	75	—	—	15	85	2750

Example Properties span the Composition, Hardness, and Transverse Rupture Strength columns.

[a] Composition and properties are averages from several manufacturers.

FIGURE 13.5
Cemented carbides are available in different grades that vary in carbide type (compositions), size, binder fraction, binder composition, processing, etc. This table shows a grading system often used in the United States. There are also ISO grades and most manufacturers have their own grading system.

Plastics

There are many plastics that can be used for unlubricated sliding in contact with hard or soft steels:

- PTFE-filled acetals
- Lube-filled polyamides (nylons)
- UHMWPE
- Polyimides + lube
- PEEK + lube
- Fluorocarbons

The fluorocarbons and some ethylene-based plastics can be used without internal lubrication, but most plastics need additives for lubrication. The original fluorocarbon, polytetrafluoroethylene (PTFE), is used in acetals and others as the lubricant. Sometimes intercalative lubricants, such as molybdenum disulfide and graphites, are used. Sometimes liquid lubricants like silicones are added during melting for injection molding. With the proper lubricant, plastics are widely used for tribocomponents.

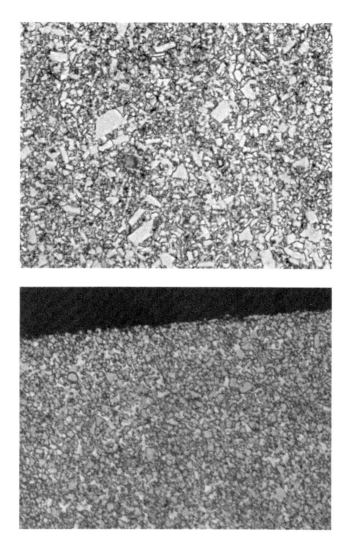

FIGURE 13.6
Fine-grain 6% cobalt cemented carbide from two different manufacturers. They may wear differently. Microstructure control is important. Both photos are magnified 1500 times.

Many plastics are reinforced with fibers of other materials to enhance various properties. When glass or carbon fibers or other hard materials are used as a reinforcement, the reinforced plastic can be abrasive to metal counterfaces. Most plastics have poor abrasion resistance (Figure 13.8), but some of the elastomers can have an abrasion resistance near that of some metals (Figure 13.9).

Elastomers tend to deform rather than cut in abrasion applications. The higher durometer ployurethanes (90–95 Shore A) are the most used for abrasion applications. Needless to say, rubbers do not work in dry sliding. Their friction is too high. However, in water, polyurethanes are often lower friction and have applications in slurry handling and pumps. Tire rubbers are the gold standard for abrasion resistance from paving materials. Styrene-butadiene rubber (SBR) is commonly used. No other material has matched the abrasion resistance of rubbers for vehicle tires.

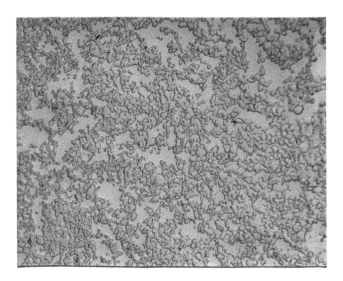

FIGURE 13.7
A cermet with a steel binder and titanium carbide (TiC) hard phase ("raised particles"). The magnification is 400 times.

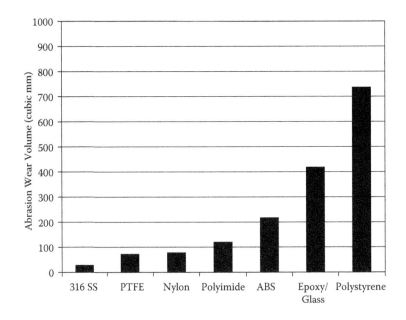

FIGURE 13.8
Abrasion wear volume of selected plastics (lower is better) compared with soft stainless steel in a modified ASTM G 65 sand abrasion test (200 revolutions).

FIGURE 13.9
Wear of an elastomer (on right) compared with hardened tool steel (D2) after an ASTM G 65 abrasion test.

Composites

The most widely used composites are fiber-reinforced plastics and thermosetting plastic laminates. These composites are not widely used for their abrasion applications, but the phenolic-cloth composites that are widely used for electrical applications have been used for decades for lubricated plain bearings. The cloth reinforcement tends to hold the oil and make it self-lubricating. Composites can be made from any of the engineering materials that we mentioned, and many new varieties come into commercial reality each year. They can be candidates for tribocomponents, but testing is probably necessary to determine how they compare with traditional tribomaterials.

Summary

Many engineering materials have tribological properties that allow their use without surface treatments or coatings. Some may require heat treatment; some must be molded to shape; some composites may require sophisticated shaping processes. There are advantages and disadvantages to each material, and this must be considered in the selection process.

Related Reading

Ashby, M., Shorecliff, H., Cebon, D., *Materials Engineering Science, Processing and Design*, Oxford: Elsevier, 2007.
Bartenev, G.M., Lavrentev, V.V., *Friction and Wear of Polymers*, Amsterdam: Elsevier, 1981.

Budinski, K.G., Budinski, M.K., *Engineering Materials: Properties and Selection*, 9th ed., Upper Saddle River, NJ: Pearson Education, 2007.

Callister, W.D., *Material Science of Engineering*, 7th ed., New York: John Wiley, 2006.

Ductile Iron Handbook, Materials Park, OH: ASM International, 2010.

Friedrich, K., Schlarb, A.K., *Tribology of Polymeric Nanocomposites*, Amsterdam: Elsevier, 2011.

Glaeser, W.A., *Materials for Tribology*, Amsterdam: Elsevier, 1992.

Laird, G., Gundlach, R., Rohrig, K., *Abrasion-Resistant Cast Iron Handbook*, Materials Park, OH: ASM International, 2007.

Yamaguchi, Y., *Tribology of Plastic Materials*, Amsterdam: Elsevier, 1990.

Zum Gahr, K.H., *Microstructure and Wear of Materials*, Amsterdam: Elsevier, 1987.

14

Surface Engineering Processes and Materials

Surface engineering is the use of coatings and treatments to improve selected properties of functional surfaces of solids. It includes all coatings and all treatments to surfaces. The processes range from covering surfaces with nano-sized surface texture features to fusion hardfacing with welding techniques. The surface engineering processes that have significant applicability for tribocomponents are shown in Figure 14.1. This chapter will describe those processes as they may be used for tribocomponents to solve friction wear and erosion problems.

Heat Treating Processes

In Chapter 13, we discussed through hardening of metals to make them suitable for wear and erosion applications, but many times only the functional surface needs to be improved and a variety of heat treating processes (Figure 14.1) can be employed to produce usually higher surface hardness. With appropriate materials, such as 1040 and 1060 steels and certain cast irons (Table 14.1), selective hardening can produce hardened depths from 0.25 to over 1 mm. Selective hardening is simply using a heating process to raise a local area or just the surface of a steel to its hardening temperature and allowing the mass of the material to provide the necessary fast quenching rate, or the use of water cooling and the like. Localized heating can be done by oxy-fuel torches, induction, laser, or electron beam (Figure 14.2). Each process has advantages and disadvantages. However, a significant advantage of selective hardening processes over other surface hardening processes is that very large parts, like 3-m diameter bearing races, can be hardened with less risk of distortion than with processes that heat the whole part.

The diffusion processes harden by elevated temperature diffusion of selected elements into a material to alter surface properties. The processes listed in Figure 14.1 apply mostly to low carbon or alloy steels, but some can be used on tool steels to enhance surface hardness. All diffusion processes are governed by Fick's laws of diffusion:

$$dx/dt = -D \, dc/dt$$

where:
- x = diffusion distance
- t = time interval allowed for diffusion
- c = concentration of the diffusing species
- D = diffusion coefficient, a constant for a particular atomic species; it is temperature dependent

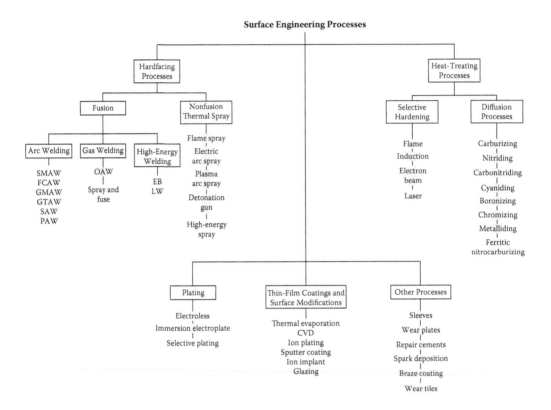

FIGURE 14.1
The spectrum of surface engineering processes.

The important diffusing species for tribological applications include:

C: carburizing

N: nitriding

B: boronizing

C + N: carbonitriding, cyaniding, ferritic nitrocarburizing, etc.

There are special processes that diffuse chromium, aluminum, and many other elements, but carbon and nitrogen are the workhorses of diffusion treatment. The depth of the diffusion can be 0.1 mm or less to 1.5 mm or more.

Each process has advantages and disadvantages (Figure 14.2). Some produce mostly thin hardened depths; some processes harden deep. Applicable steels for carburizing using the most widely used processes are low-carbon steels and low-carbon alloy steels. Process temperatures range from 482 to 500°C for nitriding, 760 to 870°C for carbonitriding, and 815 to 1090°C for carburizing.

Many times diffusion processes are used for cost reasons. A low-cost steel can be transformed into a hard surface. However, some diffusion processes can enhance the surface hardness of hardened tool steels. For example, nitriding of H-13 tool steel plastic mold makes it resistant to scratching and dings from part removal. Nitriding of special steels can produce surface hardnesses of 70 HRC, harder than any other steel. So, there are wear and erosion advantages of diffusion processes. Thus, diffusion processes are a significant part of surface engineering and will continue in that capacity for the foreseeable future.

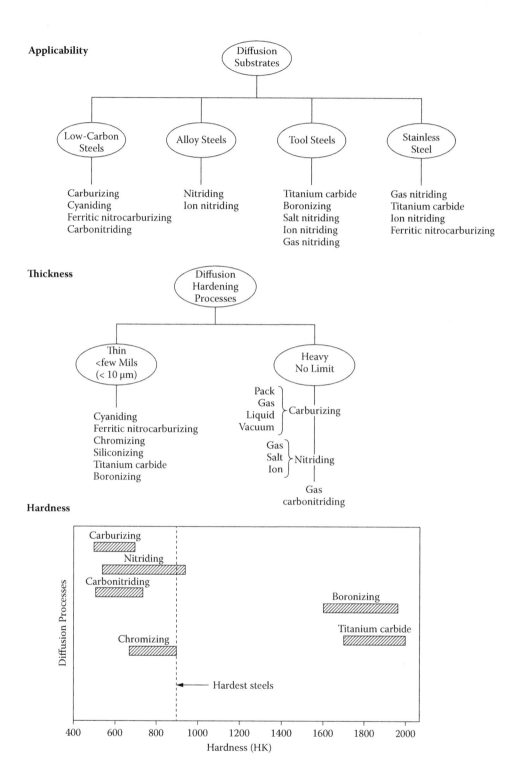

FIGURE 14.2
Comparison of diffusion heat treatments.

TABLE 14.1

Materials Commonly Selectively Hardened by Flame, Induction, Electron Beam, or Laser

Carbon Steels (HRC)		Alloy Steels (HRC)		Tool Steels (HRC)		Cast Irons (HRC)	
1025–1030	(40–45)	3140	(50–60)	01	(58–60)	Meehanite GA	(55–62)
1035–1040	(45–50)	4140	(50–60)	02	(56–60)	Ductile, 80-60-03	(55–62)
1045	(52–55)	4340	(54–60)	Sl	(50–55)	Gray	(45–55)
1050	(55–61)	6145	(54–62)	P20	(45–50)		
1145	(52–55)	52100	(58–62)				
1065	(60–63)						

Plating Processes

Plating shops usually do all of the processes shown in Figure 14.3. They do electroplating of metals and plastics; they do electroless plating of metals on metals and nonmetals, and they usually do electrochemical conversion coating on aluminum. We know these latter processes as anodizing and hardcoating. Plating is a very useful surface engineering process because special purpose metals can be applied to functional services in thicknesses ranging from a few micrometers to as much as a millimeter. Properties can range from the oxidation resistance of gold to the abrasion resistance of chromium. Electroplating can be done by both immersion and localized (selective) plating with a cloth-covered carbon electrode that floods the work with pumped plating solution. Selective plating is used on parts too large or too inconvenient to immerse in a plating tank. What makes plating adhere to a surface? The deposit metals are not diffused and they are not fused. They have a mechanical-atomic bond produced by atomically cleaning surfaces with pretreatments before plating. The plating deposit is added to the surface one atom at a time. When different atomic species are in atomic contact, there can be adhesive forces that are very strong; this is what holds platings to a surface.

Chromium is the most popular plating to be used on metals for abrasion resistance. It is a soft metal; however, when it is electrodeposited it forms a nanostructure that makes it the hardest metal at 70 HRC, and it is also corrosion resistant. It has been used for decades for all types of wear, erosion, and tribocorrosion applications. Chromium adheres best to low-carbon and -alloy steels, but most steels can be chromium plated. Hydrogen embrittlement can be a problem with hard steels. They can be prone to cracking, and procedures are needed to remove hydrogen pickup during electroplating.

Figure 14.3 lists the electroplatings that have utility in wear applications. Heavy, hard chromium is used for abrasion resistance, for example, in handling roofing shingles. Thin dense chromium has the bond and low surface roughness to allow its use on rollers and balls for rolling element bearings. Proprietary nodular chromium is done in specialty plating houses, and its pebble-like surface (Figure 14.4) can help hold lubricant for use under boundary lubrication conditions. There are proprietary chromium platings that can be used on aluminum. Most plating shops do not offer chromium plating on aluminum. Finally, silver electroplating can be used to prevent galling and seizure on mating parts, such as fasteners that are used at elevated temperatures and need to be removed afterwards.

A fundamental problem with the use of electroplating for wear applications is edge buildup. As shown in Figure 14.5, electrodeposits are at least twice as thick on edges

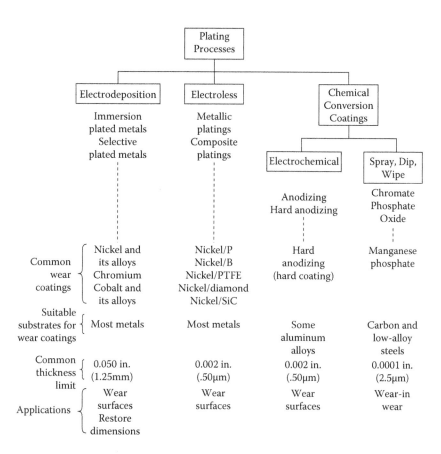

FIGURE 14.3
The spectrum of plating processes that apply to tribological applications.

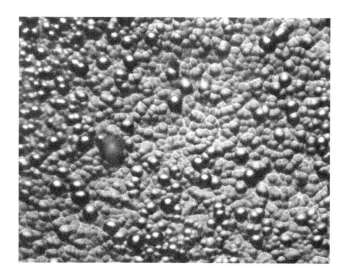

FIGURE 14.4
Nodular chromium (1100×).

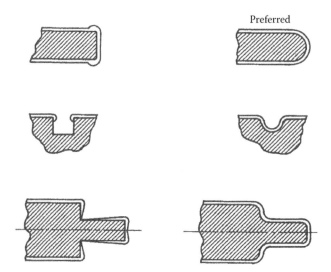

FIGURE 14.5
Edge buildup in electroplating.

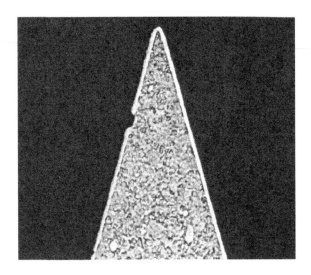

FIGURE 14.6
Edge buildup dealt with by thin deposits: Cr (2 μm thick) on knife edge (800×).

as they are away from edges, as illustrated in Figure 14.5. This problem is dealt with by radiusing edges and the use of thin deposits (Figure 14.6).

Another problem using electrodeposited chromium is the environmental issues with the disposal of hexavalent chromium plating solutions. Regulations on disposal raise costs significantly. Nonetheless, chromium electroplating is a great tool for many abrasive wear applications. However, it does not like to rub on itself. Self-mated sliding should be avoided. Chromium-plated stainless steel often is a low-cost solution for liquid erosion/cavitation problems. The piston rings used in many internal combustion engines have been plated with thin, hard chromium for decades. It performs well in contact with cast iron in lubricated conditions.

Electroless plating can be done with a variety of metals (nickel, copper, silver), but electroless nickel is the most useful from the wear and erosion standpoint. The term *electroless* has been used for this process for decades, but the correct term is *autocatalytic plating*. The metal is deposited by immersing the part in a hot aqueous solution (about 100°C) containing metal salts, a reducing agent, and other chemicals that control solution pH and reaction rates. With a suitable substrate (most metals and many nonmetals), the initial deposit acts as a catalyst to cause ions in solution in the bath to be reduced to metal atoms by the reducing agent. The ions are not picking up electrons from the work as in electroplating. The reducing agent is causing the metal ion reduction. The nickel deposited assists further deposition. The plating does not stop when metal completely covers the surface. The most common electroless nickel plating baths contain phosphorus (from the sodium hypophosphite reducing agent), and thus electroless nickel deposits can contain up to 13% phosphorus. The as-deposited hardness is usually about 43 HRC, but thermal aging treatments at about 300°C can cause deposit hardnesses to reach 60 HRC. Normal thicknesses are up to 50 μm, and plating rates are about 10 μm/h. As deposited, these coatings have successfully reduced lubricated sliding wear, but they are not very abrasion resistant even at 60 HRC.

A feature of the electroless nickel process is the ability to co-deposit other materials with the nickel by suspending solid particles in the plating solution. For example, composite platings can be produced containing diamond and other hard particles for abrasion applications (Figure 14.7), or lubricious particles, such as intercalative compounds, can be co-deposited for sliding wear applications. They reduce wear and erosion in some applications, but testing for a specific application is usually necessary.

Chemical conversion coatings are established surface treatments for many applications. Some processes are intended exclusively for tribological applications. There are electrochemical processes, and some are simple immersion processes. The most important electrochemical processes, anodizing and hardcoating, are performed mostly on aluminum. Aluminum anodizing is usually performed like an electroplating operation, except that the part is made the anode in a suitable bath like 10% sulfuric acid. The cathode is stainless steel or lead. As the voltage increases in the cell, aluminum dissolves and a reaction product is formed that is essentially aluminum oxide, a hard ceramic. The coating will start to

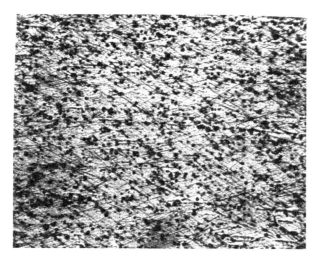

FIGURE 14.7
Electroless nickel containing 15 μm diamond particles (50×).

dissolve when attempts are made to make the coating thick, but shops offer two varieties of anodizing: anodizing and hardcoating. Chemically they are both aluminum oxide with a hardness that may be as much as 1000 kg/mm². The usual thickness limit of anodizing is 25 µm, and it is used mostly for decorative applications. Hardcoating normally has a thickness of 50 µm, but some platers can reach 150 µm. The bath temperature is controlled (chilled) to allow hardcoating thicknesses.

Anodized coatings are formed from the part's surface so it loses dimension, as shown in Figure 14.8. Not all aluminum alloys can be hardcoated. Silicon is the alloy element that makes aluminum alloys unsuitable for hardcoating. Some alloys that are suitable are shown in Table 14.2. Other metals can be anodized, titanium, magnesium, etc., but the coatings are usually not as thick or hard as those produced by aluminum anodizing. However, a plasma-assisted anodizing became commercially available in about 2000, and it produces thick (150 µm) and hard (greater than 1000 kg/mm²) coatings on aluminum and magnesium. In spite of its hardness, hardcoating is usually not found to be as abrasion resistant as hard chromium (Figure 14.9). This is probably due to the porosity that naturally occurs in anodized coatings.

The chemical conversion coatings listed in Figure 14.3 vary in thickness from a few micrometers to as much as 25 µm. Some, like zinc phosphate, are used for corrosion resistance, but the manganese phosphate chemical conversion coatings have great utility in

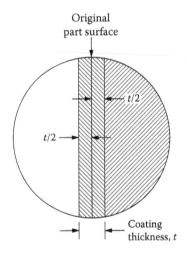

FIGURE 14.8
Formation of hardcoating from an aluminum surface.

TABLE 14.2

Suitable Aluminum Alloys for Hardcoating

Preferred Alloys for Hard Anodizing	Difficult Alloys
5052	2011
5050	2017
6061	2024
6063	7075
3003	Cast and wrought alloys with
1100	Cu > 4% or Si > 7%

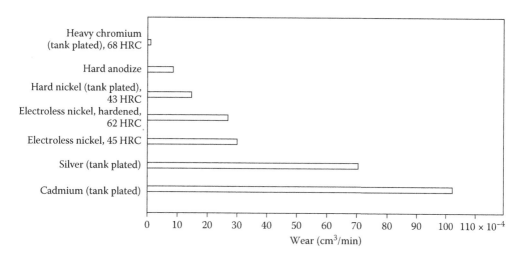

FIGURE 14.9
Abrasion resistance of various platings.

assisting break-in wear in lubricated sliding systems. These coatings are chemical reaction products produced by simple immersion in liquid baths; they are corrosion products. Since they are formed from the part surface, they have significant adhesion—they do not spall. They are soft and porous, thus making them compliant and accommodating to lubricants. Manganese phosphate coatings are particularly useful for sliding wear applications.

Thin-Film Coatings

There is no agreed-to definition of what constitutes a thin film. In the 1980s, it was 1 to 2 µm, but since the "nano craze" of the new millennium, thin-film coatings are probably coatings in the range of 10 nm to 1 mm, with less than 2 µm being the norm. Figure 14.1 illustrates the types of coatings covered by our definition of thin-film coatings: chemical vapor deposition (CVD) and coatings applied by vacuum by physical vapor deposition (PVD) technologies. Most are less than 2 µm in thickness, but some jet engine PVD coatings are applied very thick, approaching 1 mm. Two basic processes are used the most: PVD and CVD. The former involves heating a material to be coated in a vacuum until it forms a vapor that condenses on workpieces to form a coating. Chemical vapor deposition uses chemical reactions to form vapors that can condense on workpieces to form a coating. There are as many processes for producing these coatings as there are people offering coatings. Almost every supplier has its own process. In physical vapor deposition, the common denominator is the use of a vacuum chamber. Coatings are built atom by atom, and there are many processes for urging materials into atomic form. Thermal evaporation involves heating the material to be coated in a refractory metal boat until it forms a gas or vapor, and this gas produces the coating. Typical materials that can be coated are shown in Table 14.3.

Sputtering uses atom bombardment from a plasma field to dislodge atoms or molecules from a target of coating material. The liberated atoms/ions are attracted to the work to form the coating. Direct-current processes are used when both work and target are electrical

TABLE 14.3

Materials Coated by PVD and Related Processes

Pure Metals	Precious Metals
Aluminum	Platinum
Iron	Palladium
Cobalt	Rhodium
Copper	Gold
Nickel	Silver
Cadmium	**Alloys**
Silicon	Brass
Germanium	Inconel
Tin	MCrAlYs
Refractory Metals	**Ceramics (Metal Compounds)**
Chromium	Silicon oxide
Tungsten	Tantalum oxide
Molybdenum	Titanium oxide
Tantalum	Aluminum oxide
	Magnesium fluoride

conductors, and radio frequency (RF) plasma processes are used when nonconductors are involved. Sputtering can apply pure metals, alloys, inorganic compounds, and even some polymeric materials.

Ion plating is a form of physical vapor deposition in which coating atoms are converted to ions, and then they can impinge on the substrate to form coatings with enhanced bond. Table 14.4 lists some of the materials that can be coated. Ion plating is mostly used for metallic coatings that can be produced by PVD thermal evaporation. Sputtering is often used to apply compounds. From the wear and erosion standpoint, titanium nitride, chromium nitride, aluminum nitride, and titanium aluminum nitride are offered by most coating suppliers. Sometimes they are mixed to produce graded coatings. They are usually applied 1 to 2 μm thick, and they are applied mostly to improve abrasion resistance. A problem exists with all the vacuum coatings that are arc discharge generated: the formation of microscopic spheres of the hardcoatings produced by when arcs form (Figure 14.10).

TABLE 14.4

Thin-Film Coatings for Tribological Applications

Thermal Evaporation	Sputtering	Ion Plating
Au	SiO	Cr
Ag	SiO_2	Mo
MCrAlYs	Cr	TiC
Cr	Mo	TiN
Mo	Au	Au
	TiC	Ag
	TiN	Si_3N_4
	Al_2O_3	
	WS_2	
	MoS_2	
	Si_3N_4	
	PTFE	
	TiB_2	

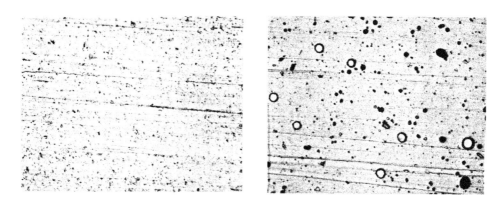

FIGURE 14.10
TiN macros produced by arc cathode sputtering process. Before (left) and after (right) coating (800×).

These spheres stand proud on the surface and tend to act as files on anything that rubs on the coated service. Most coating suppliers have learned how to eliminate these defects, but users must query suppliers about their capabilities.

Figure 14.11 shows the chemical vapor deposition processes. With the advent of diamond-like coatings there are now many more options. CVD processes may not need a vacuum chamber, but some do. Essentially these processes expose the surface to be coated to a chemical species that reacts with the surface to produce a coating. For example,

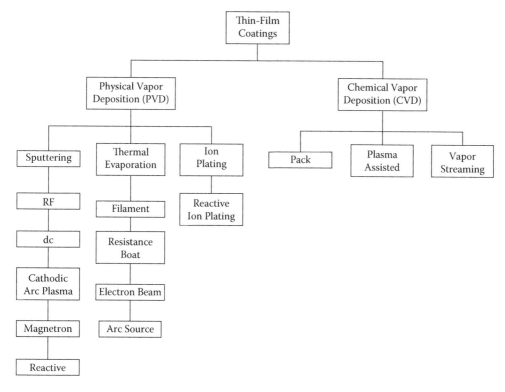

FIGURE 14.11
The spectrum of thin-film coatings.

chemical compounds can be put in a container with parts and heated to an elevated temperature (sometimes 1000°C), and gases form that react with the surface to produce a chromium- or boron-rich coating. Sometimes gases can be pumped into a chamber containing a heated part to produce a coating, or gases may be introduced into a plasma in the vacuum chamber to produce coatings.

Diamond-like carbon (DLC) coatings are commercially used for machinery parts and automobile parts. There are many types, and most have components of SP_3 carbon bond or diamond bonds in a structure. Some DLC coatings contain hydrogen, some contain metal atoms. Most are proprietary, but, in general, they can be very valuable for sliding wear applications. They slide against themselves or on another solid. Often only one surface needs to be coated because the coating will transfer to an uncoated counterface so that it rubs on itself even though only one member was coated. They can make significant improvements in wear life. However, some DLC coatings are removed by elevated temperatures (as low as 200°C). In any case, they need to be in a user's repertoire of tribological coatings.

Special Surfacing Processes

Rebuilding cements: There are thixotropic epoxies filled with hard particles that have been successfully used to repair worn parts. Figures 14.12 and 14.13 show the hard phases in one of these repair cements, and Figure 14.14 illustrates their use. They seem to work well for slurry erosion on cast iron pump casings and similar applications.

Wear tiles: Various shapes of ceramic or cemented carbide tiles can be used to line equipment as shown in Figure 14.15. They are most often used for abrasive wear applications. Successful use requires that the joints between tiles be tight.

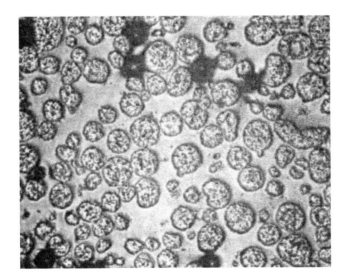

FIGURE 14.12
Epoxy trowel-on rebuilding cement containing aluminum oxide spheres (100×).

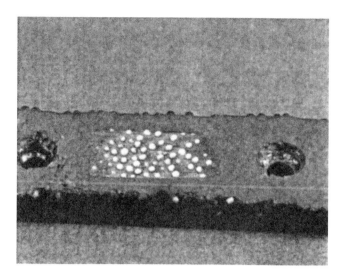

FIGURE 14.13
Alumina-filled repair cement after ASTM G 65 sand abrasion test.

FIGURE 14.14
Cast iron pump housing repaired by trowel-on repair cement.

Brazed carbide cladding: A number of commercial companies offer brazed-on tungsten carbide coatings that can be up to a millimeter thick. One supplier offers a cloth with carbide particles bonded with the polymer before forming of the cloth (Figure 14.16). The cloth can be draped on parts and furnace brazed to form a thick carbide coating.

Centrifugally cast wear coatings: Extruder barrels and the like can be coated on the inside diameter by charging the cylinder with an alloy powder material that melts at a lower temperature than the cylinder. The melted internal coating can be a white iron, a nickel–chromium–boron alloy, a cobalt-based hardfacing, or like a tungsten carbide or cermet (Figure 14.17). These coatings are standard for most plastic extrusion barrels where they must resist product abrasion and metal-to-metal wear from the rotating extruder screw rubbing on the barrel.

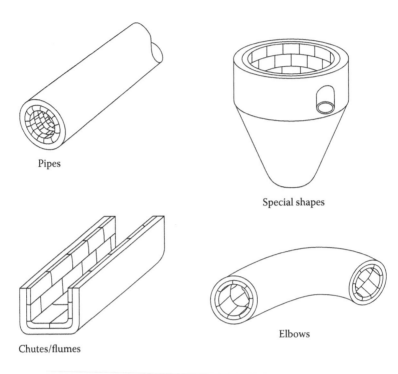

Pipes

Special shapes

Chutes/flumes

Elbows

FIGURE 14.15
Use of wear tiles to prevent abrasion and erosion.

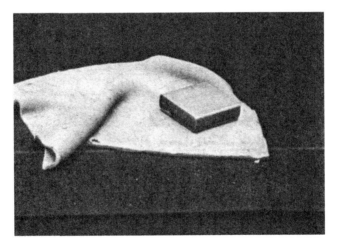

FIGURE 14.16
Carbide braze cloth that can be applied to suitable surfaces to provide a WC/Co coating that can be 500 μm thick.

> **Wear plates:** Hard white iron or other hard alloys are made into plates that can be plug welded into place, as shown in Figure 14.18. These are usually intended for abrasion applications. There are also prehardened high-strength low-alloy steel plates available, and hardnesses from 30 to 50 HRC. Most can be welded in place. Many times these products are used to resist gouging abrasion incurred in rock crushing and handling applications.

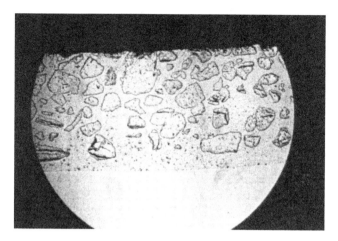

FIGURE 14.17
Cross section of a carbide composite cylinder lining (25×).

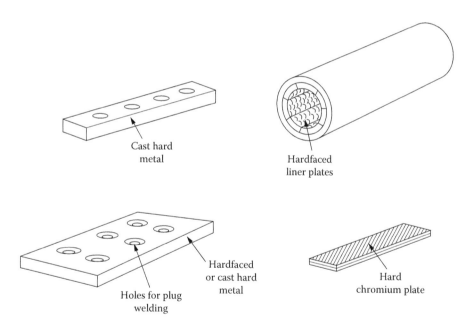

FIGURE 14.18
Use of wear plates for abrasion and erosion.

Hardfacing Processes

Hardfacing is applying to solid surfaces with welding techniques materials with properties that are superior to those of the starting solid surface. The applied materials are perceived to have wear or erosion properties that are much better than those of the substrate. For example, if the leading edge of an excavator bucket was allowed to remain as the low-carbon steel, its manufacture knows that that edge would soon be unusable. Excavator operators know that there are hardfacing overlays deposited by fusion welding processes that will protect the leading edges, and they also know that these hardfacing deposits can be refreshed when they start to abrade. Thus, they address an abrasion problem that they know with certainty exists in digging most soils.

Our definition of hardfacing separates it from other coatings with the limiting phrase, by welding techniques. Some hardfacing processes apply coatings that may be only 25 μm thick, thus making them a thin-film process, but the thin films are not applied by welding processes. On the other extreme, fusion welding deposits can be 5 or 10 mm thick—possibly qualifying them as bulk materials. However, they are applied by welding techniques, so that qualifies them as hardfacings.

There are a very large number of hardfacing processes available. They can be put into two categories: fusion and nonfusion, as shown in Figure 14.19. Fusion processes require melting of the consumable to be applied and the substrate; the nonfusion processes do not require melting of the substrate. Fusion techniques involve melting of the substrate to varying degrees. Some processes may only melt a few micrometers of the substrate, as with some laser processes, and others millimeters, as with the processes that use heavy consumables. The mixing of the consumable and the substrate in the substrate surface is called dilution, and it is a process consideration (Figure 14.20). The nonfusion process deposits adhere to the substrate by mechanical bonding; some of this bonding can be macroscopic mechanical locking, as shown in Figure 14.20. The dovetail features are created to mechanically lock the coating. Sometimes surfaces are abrasive blasted to enhance adhesion, and there are high-velocity processes that propel molten droplets of hardfacing materials at substrate surfaces with speed sufficient to clean surface films and get atoms close enough to have bonds similar to those produced by plating techniques. However, they all have mechanical bonds, and all of these techniques are termed thermal spray coatings.

Appendix I describes the major fusion and nonfusion hardfacing techniques in some detail. Each has a unique property that leads to each process having an application niche (Figure 14.21). Fusion process deposits adhere with the strength of the substrate if there is metallurgical compatibility. For example, one can overlay titanium with a shielded metal arc welded (SMAW) deposit of a cobalt-based hardfacing, but the deposits will pop off after solidification. A brittle intermetallic is formed in the dilution zone because titanium is incompatible from the solid solution standpoint with cobalt. Metallurgical phase diagrams apply to fusion welding processes. If they show that a desired consumable is not soluble in the substrate metal, it will present the cobalt-titanium type of problem: deposits may spall. Brazing processes are more forgiving than arc processes because they melt less of the substrate, and nonfusion processes can be applied to any substrate that can withstand the application temperatures. Thus, the substrate requirements for nonfusion processes are simply that they do not degrade the substrate in the application process and be prepared to meet process requirements. Fusion processes require metallurgical compatibility and sufficient thickness to not melt or significantly distort the substrate in the application process. This latter requirement is most often the factor that limits the applicability of fusion processes. It is very difficult to apply a weld deposit on a substrate with a wall thickness of 0.25 mm without risk of melting through the wall. However, a nonfusion process can be found to apply to almost any wall thickness. This is a significant factor to consider with regard to use of fusion processes versus nonfusion processes. Another factor to consider is the porosity that is common in thermal spray coatings. Porosity can be significant on the low-velocity thermal spray processes (up to 20%) to almost insignificant in high-velocity processes. However, all nonfusion processes involve propelling molten

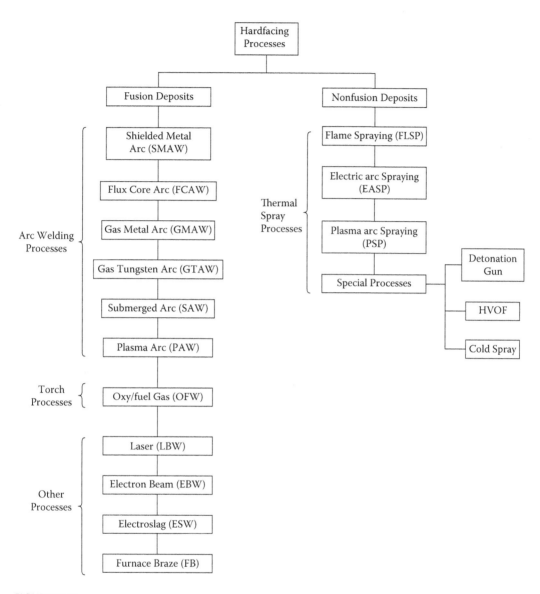

FIGURE 14.19
The spectrum of hardfacing processes.

droplets or solid particles at a solid surface, and these droplets or particles form splats that overlap each other to create a contiguous coating. Because of the nature of this coating formation mechanism, it is possible that voids (porosity) will occur when one does not conform to previously deposited splats (Figures 14.22). Figure 14.23 shows a typical nonfusion deposit on a shaft. Figure 14.24 presents some application guidelines.

What kinds of coatings are available for application by hardfacing processes? Figure 14.25 shows the spectrum of materials that can be applied by thermal spray processes. Fusion welding consumables can be identified by the American Welding Society and other national standards, for example, ANSI/AWS A5.13-10. Fusion consumables are often specified by powder supplier trade name. Figure 11.3 (Appendix II) shows the hardness

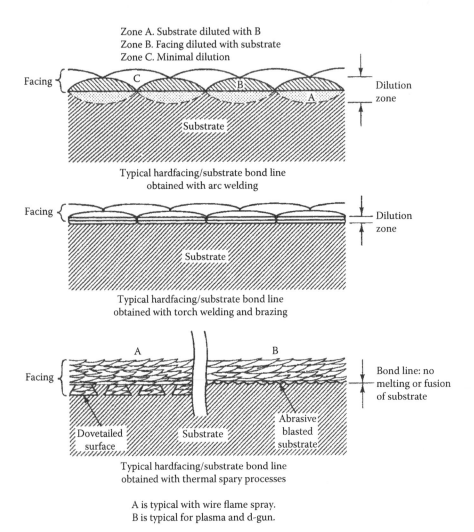

FIGURE 14.20
Bonding with arc processes, flame processes, and thermal spray processes.

ranges obtainable by widely used consumables. The E prefix means electrode; the R prefix designates an uncoated metal rod. The chemical notation designates the basis metal and major alloy components; Fe consumables are ferrous, Co designates cobalt based, etc. Figure 14.26 shows where the various consumable properties apply to different types of wear and erosion. This diagram is intended to show that the different wear and erosion processes require different material properties, and potential hardfacing users must decide on desired deposit properties before selecting a process and consumable. Hardness has the most arrows and crystal structure the least. Figure 14.27 is an example of a typical hardfacing application.

In summary, this chapter presented a spectrum of surface engineering processes and materials that can be used to address tribology problems. If desired use properties cannot be met economically with bulk materials, then surface engineering processes and materials can be employed to do the job.

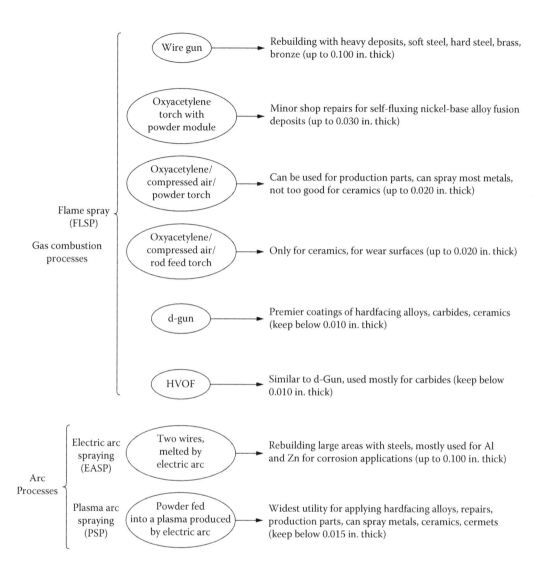

FIGURE 14.21
Applicability of thermal spray processes.

FIGURE 14.22
Porosity in a plasma-applied aluminum oxide coating (100×).

FIGURE 14.23
D-gun coating on stainless steel shafts (white is alumina; black is chromium oxide).

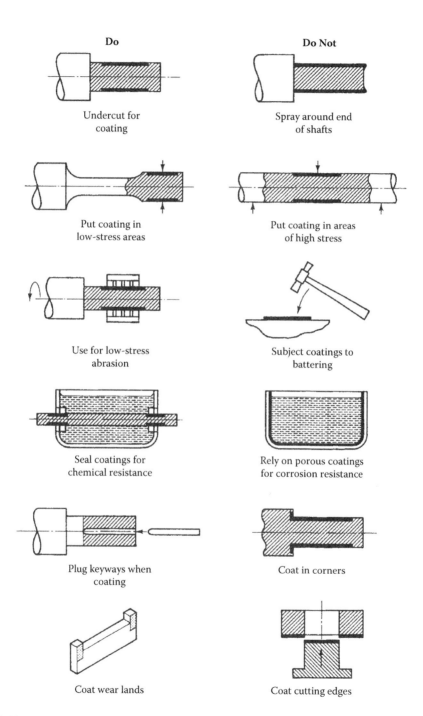

FIGURE 14.24
Application guidelines for thermal spray coatings.

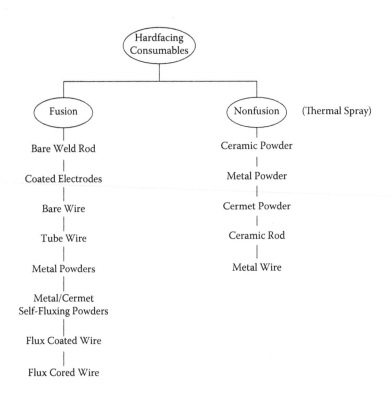

FIGURE 14.25
Hardfacing consumables.

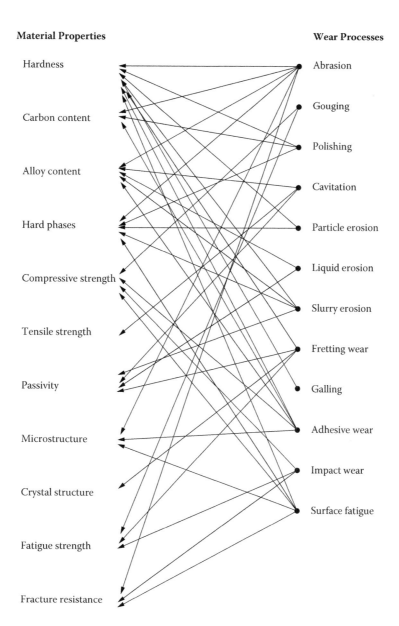

FIGURE 14.26
Hardfacing properties related to wear and erosion modes.

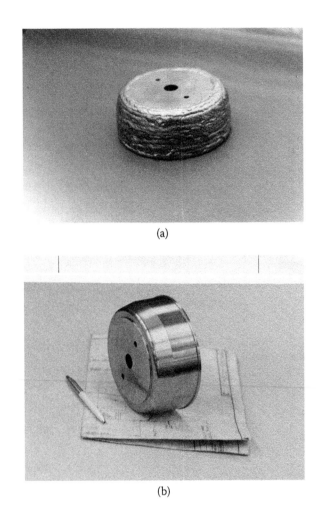

(a)

(b)

FIGURE 14.27
Typical fusion hardfaced tool. (a) as welded; (b) after finishing.

Related Reading

ASM Materials Handbook: Surface Engineering, Vol. 5, Materials Park, OH: ASM International, 1994.

Budinski, K.G., *Surface Engineering for Wear Resistance*, Columbus, OH: Prentice Hall, 1983.

Davis, J.R., Ed., *Surface Engineering for Corrosion and Wear Resistance*, Materials Park, OH: ASM International, 2001.

Holberg, K., Matthews, A., *Coatings Tribology: Properties, Mechanics, Techniques*, Amsterdam: Elsevier, 2009.

Peterson, M., Winer, W., Eds., *Wear Control Handbook*, New York: ASME, 1980.

Totten, G., Liang, H., *Surface Modifications: Friction Stress and Reactions*, Boca Raton, FL: CRC Press, 2004.

15

Wear and Erosion Solutions

The purpose of this book is to show what wear, erosion, and friction forces look like and to suggest ways to reduce their negative effects. Friction, of course, can be negative or positive, depending on what part of a vehicle or device you are referring to (bad in the engine, good in the brakes). Wear and erosion are almost always undesirable. Our objective is to help readers of this text reduce the negative costs in each reader's particular circumstance. It is the purpose of this chapter to suggest a methodology for selecting a material and process to address the negative aspects of various wear and erosion modes that were presented in early chapters. Our objective is implementation of the tribology information in this text to solve problems. We will discuss solution optimization, methodology, using bulk materials, and surface engineering and laboratory testing.

Solution Matrix

Where does one start in the solution process? There are only two kinds of wear and erosion problems: existing and perceived. If a solution is needed for either instance, the decision must be made as to what mode of wear or erosion exists. Figures 2.1 and 2.2 can help in the decision; if they are not enough help, the photos in the wear and erosion chapters (Chapters 3 to 10) may help. If a solution is needed for an existing wear problem, the nature of the part wear can be analyzed to establish a wear mode. Examine it visually, with a loupe, then with microscopes, etc., as needed. If multiple wear and erosion modes exist in the same component, list them all. This is the first step in a solutions matrix, one of which is outlined in Figure 15.1. The required tribocomponent properties are listed vertically, and the candidate material solutions make up the columns of the matrix. Numbers are established for each property for each candidate material. Candidate solutions can be quantitatively compared.

To illustrate how this technique works, it will be applied to a theoretical wear/erosion tribocomponent: a blade for a small 30 in. rotary lawnmower. We assume that this is a new product. The blade will be like that on existing mowers, but it has a new shape that reduces noise. Step 1 is to decide on applicable wear and erosion modes. It is assumed that dulling of the blade-cutting edge will be a major concern. The blade will be dulled by material removal from rubbing on grass and striking airborne particles and stones. Thus, the assumed wear mode is low-stress abrasion, coupled with limited solid particle erosion. The severity of the solid particle erosion problem could be judged by estimating the particle impingement velocity: the tip radius is 0.25 m and the typical speed is 2000 rpm; therefore, the tip speed can be 50 m/sec. This is a speed that could produce significant solid particle erosion. However, it is decided that this is not a major property factor since rotary mowers are not often operated on bare ground that might produce airborne particles.

Property	Painted 1020 Steel	Painted Carburized 1020 Steel	Painted 4140 Steel @ 47 HRC	4140 Steel @ 47 HRC + Chrome Plate	420 SS @ 52 HRC
High yield strength	1	4	9	9	10
Rust resistance	2	2	2	8	10
Low cost	10	9	8	4	1
High toughness	1	4	10	8	8
Abrasion resistance	1	8	7	10	6
Totals	15	27	36	39	35

FIGURE 15.1
Material selection matrix for solving a material selection problem (10 is best).

It is established that low-stress abrasion is the primary wear mode; it is low stress because the stress is only that of rubbing of blades of grass, which flex easily. There is also a risk of impact from hitting rocks and other unexpected things hidden in the grass, but it is well known that mower blades cannot survive these encounters unscathed. However, it is decided that the blade must have reasonable strength (better than soft steel), the stiffness of steel, high toughness, and enough hardness to get reasonable abrasion resistance. Proceeding to step 2, list desired material properties and attributes. This process starts by listing the mower's operating conditions:

- 2000 rpm
- Possibly wet grass
- Some dirt impingement
- Maximum operating temperature of 40°C

Then fine-tune the material requirements so they may resemble the following:

- Tensile strength of at least 689 MPa
- Impact strength of at least 27 J
- Elastic modulus of at least 206 GPa
- Surface hardness as high as possible
- Resistant to grass abrasion and solid particle erosion
- Capable of being flat blanked and formed
- Machinability of low-carbon steel
- Will not rust before use
- Low cost and available material of construction

It is decided that corrosion resistance is wanted, but it is felt that material lost due to rusting when not in use is not a significant concern since all lawnmowers rust underneath after use and storage. Users are conditioned to accept this as normal. The desired property list is prioritize and reduced to the following:

1. Good impact strength (no pieces fractured on impact)
2. Steel stiffness

3. Moderate yield strength

4. Acceptable abrasion resistance

5. Rust resistance (reasonable)

6. Low cost (easy fabrication)

This is the must list. If the blade strikes a foreign object like a rock, the blade cannot shatter into pieces that might hit an operator at 50 m/sec; the material must have high toughness. Steel stiffness is needed because it cannot bend in service; the steel may be heat treated to increase the yield strength, and steel has the highest elastic modulus to prevent the blade from excessive deflection during mowing heavy grass. Abrasion resistance needs to be good enough to make it through 2 years of mowing without fracturing the end of blades of grass. Rust resistance only needs to be sufficient to at least to get through the product storage at dealerships, and the material cost must be low because mower blades are a consumable.

The next step (number 4) is to formulate a material solution matrix using the property list and candidate material solutions selected from suggested bulk materials (Figure 13.1) and from surface engineering solutions (Figure 14.1). The solution matrix may now resemble Figure 15.1. A number from 1 to 10 is assigned to each property for each material: 10 is best, 1 is the lowest rating. After the quantification step, each candidate solution has a numerical assessment of how it meets a particular property need. These relative numerical ratings are judgments on the part of the person seeking the solution or on the part of others. In organizations large enough to use teams for engineering decisions, a team can be established to develop the solutions matrix. It is also common to numerically weight desired properties and reiterate the matrix with new solutions. The rating numbers for each candidate material are summed, and the winning solution is the material with the highest number, chromium plated 4140 steel in our example.

Not all wear and erosion solutions need a decision matrix. Some people establish the matrix in their mind using their knowledge of engineering materials and experience with surface engineering processes. This is how material selection is done. The matrix often shows that some perceived solutions are not as good as others. It is almost always helpful to formulate a solutions matrix. The remainder of this chapter lists solution concerns that should be reviewed during the solution formulation process.

Material Considerations

Using materials that do not need surface engineering processes to meet performance expectations may appear to be the simplest solution. However, materials that have utility in wear and erosion applications often need heat treatment to be viable for tribological applications; a concern is cost/extra handling, distortion that will occur in heat treating.

Bulk materials commonly have availability problems. For example, wrought cobalt-based alloys only come in a few shapes, and they may not be on hand at suppliers. You may need a mill run of 1000 kg in order to get the 2 kg that you need for the part in question.

The quantity of material needed to make a tribocomponent often plays a significant role in the selection process. If a production run of 100,000 lawnmower blades is planned, then a mill run is a possibility, and most any material could be considered a selection candidate. If only six prototype blades are the goal, the selection candidates must come from materials on hand at suppliers.

Every engineering material has manufacturability issues. For example, if pre-hardened steels or wrought cobalt-based alloys are selection candidates, the extra costs to machining these materials to shape must be considered.

Machining costs are almost always an issue when bulk materials are used. In the area of business issues, some candidate materials may have significant health hazard concerns. For example, beryllium copper alloys have some unique properties that make them candidates for tribocomponents, but because they contain beryllium, there are government regulations that require precautions in machining and handling. Some cobalt- and nickel-containing alloys can have adverse effects in some in vivo medical applications. So business concerns may be a selection issue.

Our material selection process suggests listing required material properties: mechanical, physical, chemical, and business issues. Sometimes desired properties significantly limit the number of candidates or add significant cost. Also, a user must question verification of desired properties. Are handbook properties adequate or are your own tests required? If so, they can add costs. An example where a physical property may be the most important is knives for plastic sheet sealing and cutoff. Cutting-edge wear is a problem, but the right material needs to have high thermal conductivity because it provides the heat to seal the plastic, so candidates must have high thermal conductivity and high hardness. There are few bulk materials that fill this bill.

In summary, using bulk materials can create selection issues that are not initially apparent. Figure 13.1 can be used to select bulk material candidates, and Figure 14.1 can be used to select surface engineering candidates.

Surface Engineering Considerations

Many surface engineering processes involve heating a substrate material during the treatment or coating process. This poses two major considerations: (1) degradation of substrate properties and (2) distortion or size change to the treated shape. If a candidate surface engineering process involves heating a part of temperatures where substrate structure changes occur or where transformations occur, as is the case with most hardened steels, softening occurs along with reduced strength. If process temperatures reach the stress relief temperature range for the substrate material, distortion is possible (Table 15.1). Figure 15.3 illustrates the possible temperature ranges that may be encountered in typical surface engineering processes. Many involve heating in the temperature range of 530 to 650°C. These temperatures can soften a steel substrate. Table 15.2 lists hardfacing considerations.

Coating processes always produce dimension changes that may affect function. However, some diffusion processes that theoretically penetrate into the substrate can also produce surface roughening, and that may not be anticipated. Many pack cementation processes do this, and the macros produced in arc discharge thin-film coating processes often create service roughening that can affect function.

Some surface engineering processes roughen part surfaces. Figure 15.4 compares typical coating thicknesses for a variety of surface engineering processes. However, some

Service Goal	Related Property
Wear resistance	Surface hardness
	Suitable microstructure
Impact resistance	Impact strength
	Fracture toughness
Rust resistance	Atmospheric corrosion resistance
High strength	Yield strength > 100 ksi (689 MPa)
High stiffness	Modulus of elasticity = 30,000 ksi (210 GPa)

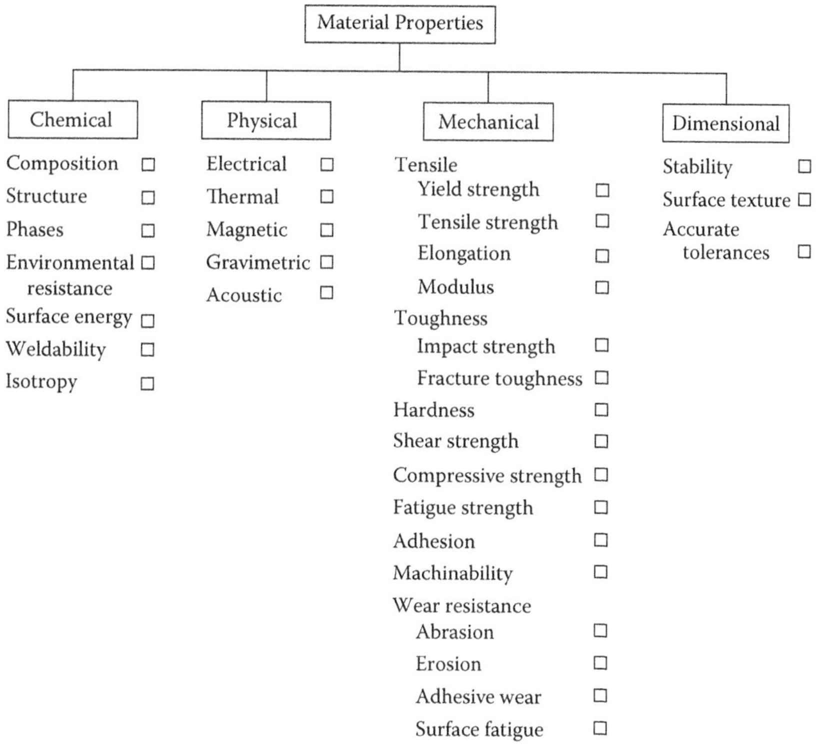

```
                    ┌─────────────────────┐
                    │  Material Properties │
                    └─────────────────────┘
```

Chemical	Physical	Mechanical	Dimensional
Composition ☐	Electrical ☐	Tensile	Stability ☐
Structure ☐	Thermal ☐	Yield strength ☐	Surface texture ☐
Phases ☐	Magnetic ☐	Tensile strength ☐	Accurate
Environmental ☐	Gravimetric ☐	Elongation ☐	tolerances ☐
resistance	Acoustic ☐	Modulus ☐	
Surface energy ☐		Toughness	
Weldability ☐		Impact strength ☐	
Isotropy ☐		Fracture toughness ☐	
		Hardness ☐	
		Shear strength ☐	
		Compressive strength ☐	
		Fatigue strength ☐	
		Adhesion ☐	
		Machinability ☐	
		Wear resistance	
		Abrasion ☐	
		Erosion ☐	
		Adhesive wear ☐	
		Surface fatigue ☐	

FIGURE 15.2
Property checklist for establishing a list of required properties.

diffusion processes like boron diffusion treatments can create a pebble-like roughening that is not part of the hardened layer. This roughening is a process-induced part change in shape. Process candidates need to be investigated for this if surface roughening can be of functional concern.

Many surface engineering processes have limitations (not size). As shown in Table 15.2, there are limitations for each process. These are for hardfacing. There is a similar list for plating, thin-film coating, thermal spray, etc.; some processes are more forgiving than others. However, any of the processes that require vacuum or a sealed chamber, like PVD and CVD coatings, pose physical size limitations. One of the biggest advantages with fusion welding hardfacing is that it can often be done in situ and on a massive piece of machinery. Size often matters. Is a part too big for a desired surface engineering process?

TABLE 15.1

Stress Relief Temperatures for Various Metals

Alloy System	Stress-Relieving Temperature (°F)	(°C)
Aluminum	300–500	148–260
Copper	400–650	204–343
Low-carbon steel	1100–1200	593–649
Cast iron	1000–1100	538–593
Alloy steel	1100–1200	593–649
Austenitic stainless steel	1600–1700	871–926
Tool steels	1100–1200	593–649
Martensitic stainless steel	1100–1200	593–649
Magnesium	300–800	148–426
Nickel	900–1300	482–705
Titanium	1000–1400	538–760

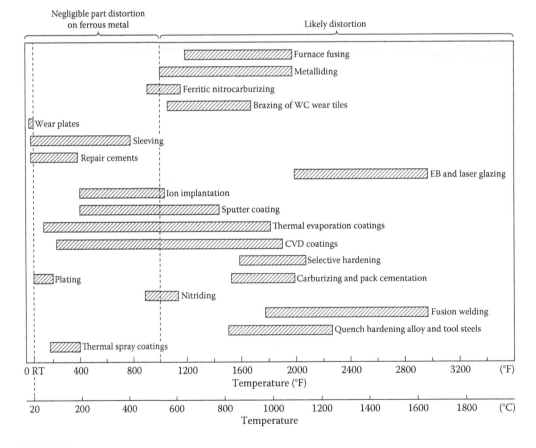

FIGURE 15.3

Maximum temperatures that can be anticipated in various surface treatment and coating processes.

TABLE 15.2

General Characteristics of Surface Engineering Processes

	Principal Limitations	Best Suited For
Fusion Processes		
SMAW	Low deposition rate	Small jobs, out of position, wide variety of consumables
FCAW	Not all alloys available	Heavy deposits
GMAW	Limited alloys available	Large jobs
GTAW	Very low deposition rate	Small jobs, tool repairs
SAW	Limited alloys, flat position	Heavy deposits
PAW	Limited alloys (must use powder or wire), expensive equipment	Mechanized production jobs
OAW	Low deposition rate	Cobalt- and nickel-based alloys, small jobs, field welding
LASER	Expensive, techniques not well developed	Special production jobs
EB	Poor availability of equipment, techniques not well developed	Special production jobs
Electroslag	Poor availability of equipment, only for alloys in bare wire form	Heavy deposits
Furnace braze	Equipment availability, only suited to a few consumables	Placement of carbide tiles
Nonfusion Processes		
Flame spraying (FLSP)	Significant porosity	Heavy deposits > 0.040 in. of metal alloys
Electric arc spraying (EASP)	May require dovetailing, only for consumables in wire form	Heavy deposits > 0.040 in. of metal alloys
Plasma arc spraying (PSP)	Expensive equipment, thin deposits < 0.040 in. (1 mm)	Cermics, cermets, thin metal alloy deposits
Detonation gun, d-gun	Proprietary equipment (one vendor)	Thin deposits < 0.005 in., 127 µm of ceramics, carbides, and high alloys
HVOF	Relatively high gas consumption	Thicker deposits (0.010 to 0.020 in., 254 to 508 µm) of tungsten carbides and high-alloy metal powders

A concern that arises in selecting a heat treatment for bulk materials or surface engineering processes is: Can the process be made continuous, or does every part need to be handled?

PVD and CVD coatings require fixturing of parts. Each part has to be handled. Many heat treatments can be done in batches. Others can be done with conveyorized furnaces. Handling adds cost, and cost is often a significant selection factor.

Some surface engineering processes like thermal spray processes produce varying degrees of porosity. This porosity may not affect function in some applications, but some applications will not be affected.

Some processes like hardfacing require listing of process details, such as the following:

- Number of passes
- Deposition process
- Surface preparation

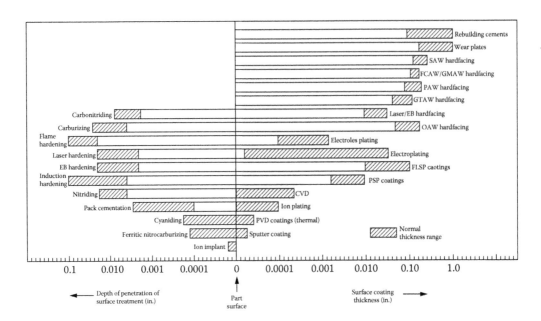

FIGURE 15.4
Typical coating thicknesses/depth of penetration for some surface engineering processes (0.02 in. = 50 μm).

- Preheat or not
- Post-heat or not
- Peening or not

Successful use of a surface engineering process often requires many process details. All coating processes involve a size change.

- Are there part surfaces that need to be masked?
- Are detailed instructions necessary for the process?
- What is the coating thickness or treatment depth (Figure 15.4)?

Laboratory Testing

Where does laboratory or prototype testing fit into the solutions process? Will a solutions matrix like Figure 15.1 have a row for ASTM G 65 abrasion resistance? Tests could be conducted on each candidate material and the data applied to the matrix. For example, wear volume data can be mathematically treated so that the wear volumes for all test materials have a quantification between 1 and 10, so they will fit into the 1 to 10 matrix rankings. In the case of the lawnmower blade, the matrix could have a data row for low-stress abrasion (per ASTM G 65) and another row for solid particle impingement results using the ASTM G 76 test. Thus, tribotesting results become a part of the material selection matrix. The key to obtaining meaningful tribotesting results is to accurately simulate the tribosystem of interest. The following is a checklist of considerations for making a tribotest simulate an application.

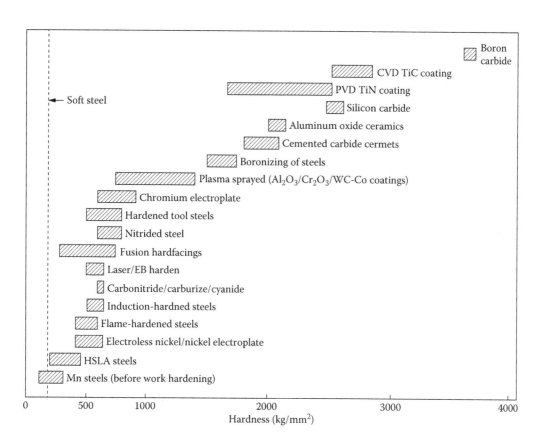

FIGURE 15.5
Hardnesses available from various material solutions.

Types of motion: Continuous ____, rolling ____, sliding ____, etc.
    Intermittent ____, reciprocating ____, oscillating ____, etc.
Contact geometry: Flat-on-flat ____, point at start ____, line at start ____, conformal ____, etc.
Sliding speed: ____
Load/force: ____
Expected life: Hours ____, 20,000 km ____, etc.
Environment: Lubricant ____, temperature ____, wet ____, humidty ____, etc.
Wear mode: Abrasion ____, galling ____, metal-on-metal ____, etc.
Erosion type: SPE ____, LDE ____, liquid ____, slurry ____, cavitation ____, etc.
    SPE: Particle type ____, size ____, velocity ____, impingement angle ____, etc.
    LDE: Droplet material ____, size ____, velocity____, duration____, etc.
    Liquid erosion: Liquid type ____, velocity ____, temperature ____, etc.
    Slurry: Slurry composition ____, particle size ____, particle hardness ____, etc.
    Cavitation: Liquid ____, field size ____, bulk liquid velocity ____, etc.

Friction:

    Type of friction: _____

    Mating couple: Smaller surface _____, larger surface _____

    Surface texture: Smaller surface _____, larger surface _____

    Apparent contact area: _____

    Relative slip: _____

    Sliding speed: _____

    System: _____

    Type of motion: _____

    Environment: _____

A checklist such as this will help to establish testing parameters and the types of tests that will truly simulate an application. That is the key. Simulate as best you can. However, do not make the test exactly like service conditions because then it becomes a life test and not a tribotest. To screen materials for an application, most tests should take no more than 24 hours. If one does, it will probably be too costly or produce too much of a delay. Accelerate the load or velocity to speed testing since these two factors are likely to be higher in service than calculations may show. Local contact stress is often orders of magnitude higher on fresh services because mating surfaces do not conform until they wear together. That is why increasing loads is valid in sliding wear tests.

Sliding velocity often must be reduced in lubricated sliding tests in order to get the rubbing surfaces to touch. Full operating speed may produce hydrodynamic film separation of surfaces. They do not touch; they will not wear; there will be no data or comparisons of materials. What to do? Reduce sliding speed until the friction force output shows that the surfaces are contacting each other. Run sliding wear tests at this speed or lower. The rationale for this is that in service all lubricated machines for devices see surface-to-surface contact in tribocomponents at start-up and shutdown, and this is the only time that wear will occur. Many lubricated wear tests use increasing loads until seizure occurs. Such tests become an evaluation of the lubricant as a machining or cutting fluid. When contact pressures exceed the compressive strength of either contacting material, the validity of the test as representative of service must be questioned. In erosion studies, it is common to concentrate erosion species on a small spot. This is analogous to increased loads in sliding tests. If friction is a concern for wearing surfaces that may be separated by wear debris, then friction tests can be run for long times and friction forces should be watched for stabilization. Use these stabilized data.

Summary

This book is intended for people, whatever their field, who have a tribology problem or perceive one in a design that is in the planning stage. Our objective is to give guidance in addressing the tribology and tribology solutions. The photos of worn and eroded surfaces are intended to help users of this book recognize the modes of friction wear and erosion that may be present in their tribosystem. The graphs and line drawings and methodology discussions are intended to assist users in selecting materials and processes to solve tribology problems. Figure 15.1 is an attempt to relate matrix solutions to modes of wear and erosion. Figure 15.5 shows hardnesses possible with various solutions because hardness is an important tool in solving tribology problems. Figure 15.3 shows process temperatures because many tribocomponents have a maximum allowable temperature that they can withstand. The typical friction force traces and Chapters 11 and 12 are intended to help users interpret friction results from tribotests and service audits. Our discussions of tribotests are not intended to tell users how to run them, just provide guidance on the selection of a valid test, one that accurately simulates the tribosystem of interest.

The overall message of this text is that friction, wear, and erosion are part of every machine, every device, and every mechanism that involves rubbing/relative motion of solids or moving contact by particles or liquids. They must be dealt with. The use of a solution matrix like the one presented on the mower blade can be a helpful tool. It forces a designer to determine all desired properties and allows quantification of them in comparing candidate solutions. This tool has been successfully used for many years.

Solving tribological problems is often not easy, but it can be done. In the 1940s, automobile engines often needed rebuilding after 50,000 miles. Engine lives of 150,000 miles are now common. Just 30 years ago, automobile tires seldom lasted more than 10,000 miles. In 2012, 50,000 miles of life was typical on even factory-issued tires. Improved materials, treatments, and coatings have allowed these gains. However, solving tribological problems starts with identifying the culprit. Hopefully this atlas will do this.

Related Reading

ASTM G 163, *Standard Guide for Digital Data Acquisition on Wear and Friction Tests*, West Conshohocken, PA: ASTM International.

ASTM G 190, *Standard Guide for Developing and Selecting Wear Tests*, West Conshohocken, PA: ASTM International.

Budinski, K.G., Budinski, M.K., *Engineering Materials: Properties and Selection*, 9th ed., Upper Saddle River, NJ: Pearson Education, 2007.

Dieter, G.E., *ASM Handbook: Materials Selection and Design*, Vol. 20, Materials Park, OH: ASM International, 1990.

Roberts, G., Kravin, G., Kennedy, R., *Tool Steels*, 5th ed., Materials Park, OH: ASM International, 1998.

Stolarski, T.A., *Tribology in Machine Design*, Oxford: Heinemann, 1990.

Summers-Smith, J.D., *A Tribology Casebook*, London: Mechanical Engineering Publications, 1997.

Appendix I: Fusion and Thermal Spray Hardfacing Processes

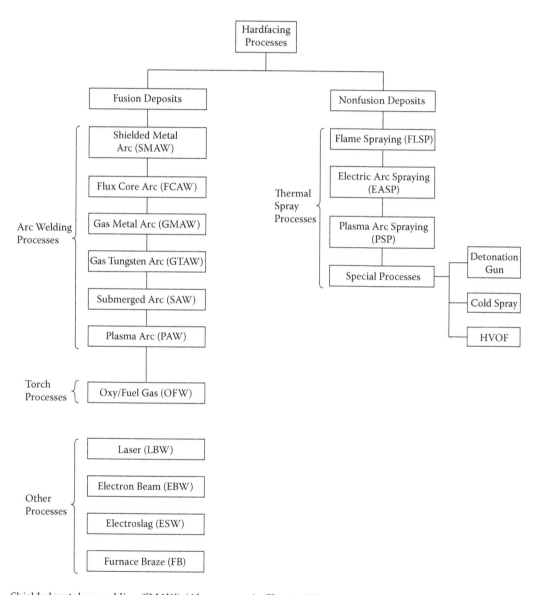

Shielded metal arc welding (SMAW). (Also appears in Chapter 13.)

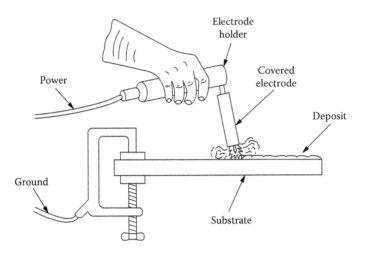

FIGURE I.1
Shielded metal arc welding (SMAW).

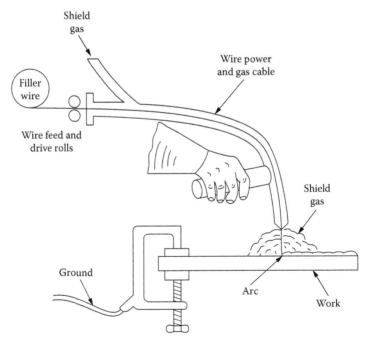

FIGURE I.2
Gas metal arc welding (GMAW).

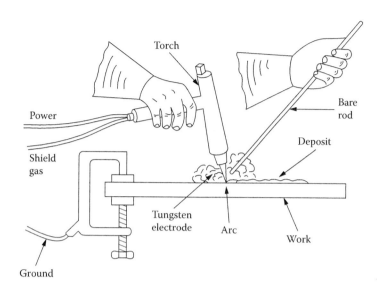

FIGURE I.3
Gas tungsten arc welding (GTAW).

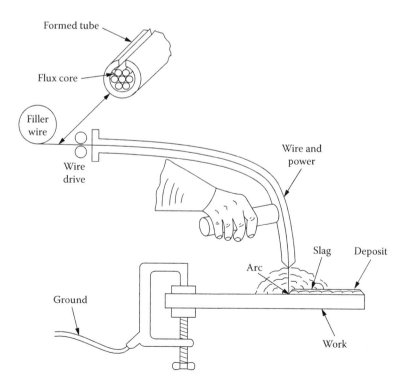

FIGURE I.4
Flux cored arc welding (FCAW).

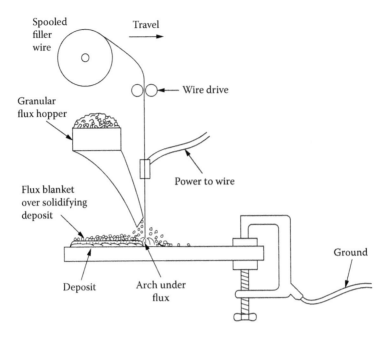

FIGURE I.5
Submerged arc welding (SAW).

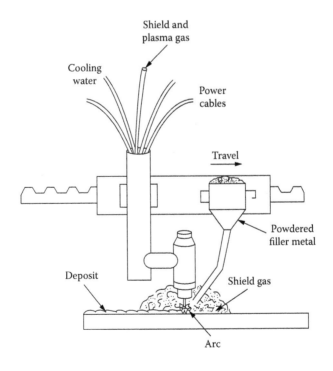

FIGURE I.6
Plasma arc welding (PAW).

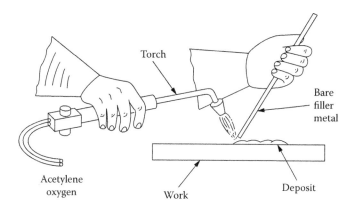

FIGURE I.7
Oxy-acetylene welding (OAW).

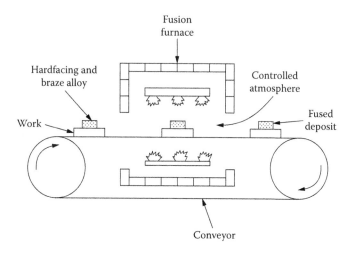

FIGURE I.8
Furnace fusing of a hardfacing consumable.

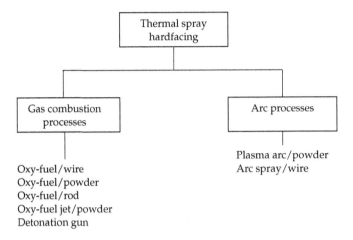

FIGURE I.9
Thermal spray processes.

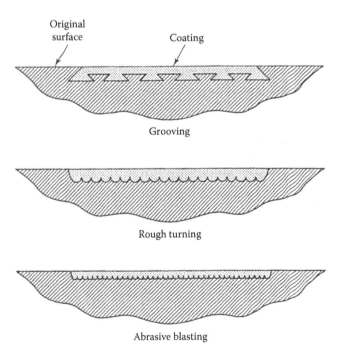

FIGURE I.10
Grooving for flame spraying.

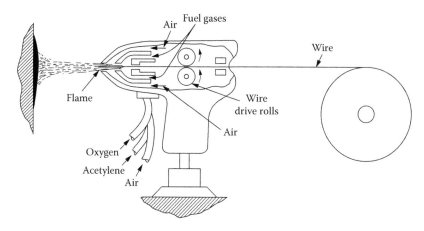

FIGURE I.11
Flame spraying a roll.

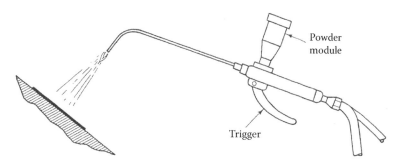

FIGURE I.12
Powder spray oxy-fuel torch.

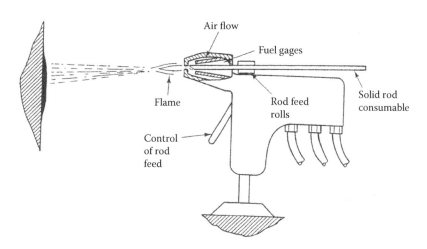

FIGURE I.13
Flame spraying with a rod consumable.

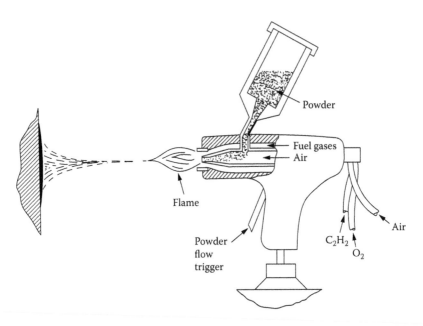

FIGURE I.14
Thermal spraying with a powder-feed gas combustion gun.

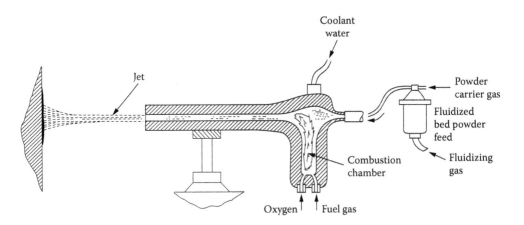

FIGURE I.15
Combustion jet hardfacing (HVOF).

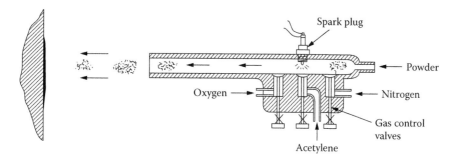

FIGURE I.16
D-gun thermal spraying.

Appendix II: Fusion Hardfacing Consumables and Design Aides

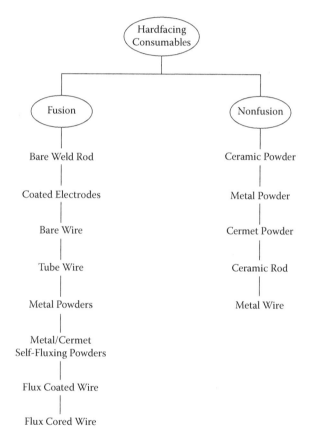

FIGURE II.1
Categories of hardfacing consumables (Appendix II deals with the first column).

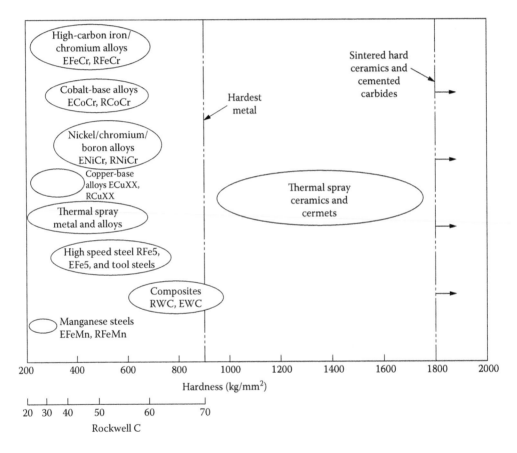

FIGURE II.2
Hardnesses available with various hardfacing consumables.

Commonly Used Fusion Welding Consumables

Type	Composition Range	Hardness Range (HRC)
Tool steels, oil hard	1/1.5 C, 0/1 Mn, 0.5/1 Cr, bal. Fe	40–55
Air hard	1/2 C, 1/1.5 Mo, 1/12 Cr, bal. Fe	40–60
Alloy steels, 4130, etc.	0.3 C, 0.9/0.6 Mn, 0.8/1.1 Cr, 0.15/0.25 Mo, bal. Fe	20–40
High-speed steels (E/R Fe5)	0.5/1.1 C, 0.5 Mn, 1/7 W, 315 Cr, 4/9 Mo, 1/2.5 V, 0.5/Si, bal. Fe	50–60 + carbides
Austenitic manganese steels (E/R FeMn)		
Nickel type	0.5/1 C, 11/16 Mn, 2.75/6 Ni, 0.5 Cr, 1.3 Si, bal. Fe	
Moly type	0.5/1 C, 11/16 Mn, 0.6/1.4 Mo, 0.3/1.3 Si, bal. Fe	20–25
Low-chromium martensitic iron	3.5/4 C, 1 Mn, 4/15 Cr, 2/4 Mo, bal. Fe	50–60 + carbides
Low-chrominum austenitic iron	2/4 C, 2/2.5 Mn, 12/16 Cr, 0/8 Mo, bal. Fe	50–60 + carbides
High-chromium martensitic iron	2.5/4.5 C, 1/1.5 Mn, 26/30 Cr, bal. Fe	40–60 + carbides
High-chromium austenitic iron (E/R FeCr)	3/5 C, 2/8 Mn, 26/35 Cr, bal. Fe	40–60 + carbides
Low-carbon nickel/chromium/ boron (E/R NiCr-A)	0.3/6 C, 1.5 Co, 8/14 Cr, 1.25/3.25 Fe, 1.25/3.25 Si, 2/3 B, bal. Ni	30–40
Medium-carbon nickel/chromium/ boron E/R NiCr-B)	0.4/0.8 C, 1.25 Co, 10/16 Cr, 3/5 Fe, 3/5 Si, 2/4 B, bal. Ni	45–50
High-carbon nickel/chromium/ boron (E/R NiCr-C)	0.5/1 C, 1 Co, 12/18 Cr, 3.5/5.5 Fe, 3.5/5.5 Si, 2.5/4.5 B, bal. Ni	50–60 + carbides
Low-carbon cobalt-based alloys (E/R CoCr-A)	0.9/1.4 C, 1 Mn, 3/6 W, 3 Ni, 26/32 Cr, 1 Mo, 3 Fe, 2 Si, bal. Co	38–47
Medium-carbon cobalt-based alloys (E/R CoCr-B)	1.2/1.7 C, 1 Mn, 7/9.5 W, 3 Ni, 26/32 Cr, 1 Mo, 3 Fe, 2 Si, bal. Co	45–59
High-carbon cobalt-based alloys (E/R CoCr-C)	2/3 C, 1 Mn, 11/14 W, 3 Ni, 26/33 Cr, 1 Mo, 3 Fe, 2 Si, bal. Co	48–58 + carbides
Aluminum bronze (E/R CuAl-D)	3/5 Fe, 13/14 Al, bal. Cu	20–40
Composite tubular rods (R/E WC)	Tube: 0.1 C, 0.45 Mn, 0.3 Ni, 0.2 Cr, 0.3 Mo, bal. Fe Carbides: 3.6/4.2 C, 94 W min	50–65 + various size carbides

FIGURE II.3
Commonly used fusion hardfacing consumables.

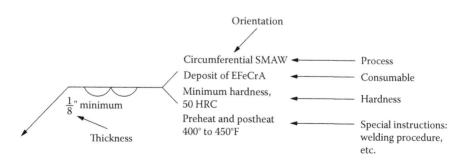

FIGURE II.4
Hardfacing weld symbol showing process and consumable details.

Deposit location:

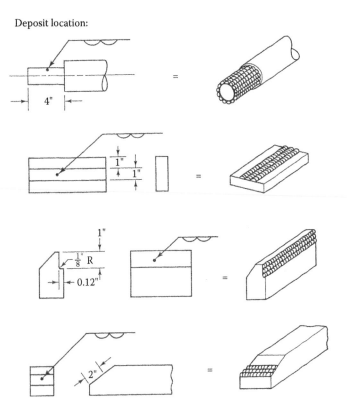

FIGURE II.5
AWS system for specifying hardfacing deposit location.

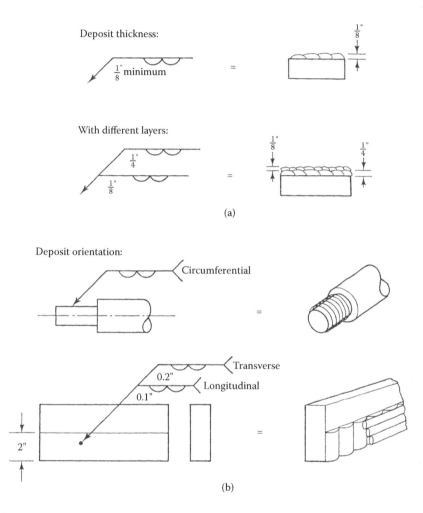

FIGURE II.6
AWS system for specifying hardfacing thickness and orientation.

Blanking die: Use tapered horizontal groove
for ease of deposition

Battering tool: Taper deposit at ends

 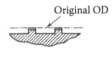

Extruder screw: Put groove in bar and deposit
surfacing before machining flights

Guide bushings: Radius weld grooves

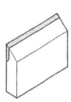

Knife edges: Prebow to reduce distortion; stress
relieve and machine angle after

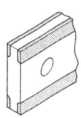

Shear blades: Welding of four edges minimizes
distortion; stagger weld

FIGURE II.7
Typical fusion hardfacing applications.

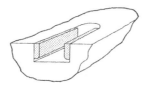

Keyway/cam follower groove: Do not weld into bottom
of groove; minimizes
cracking

Gear/sprocket teeth: Taper deposit; weld
only loaded side

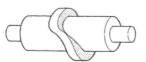 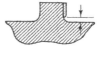

Barrel cam: Weld only loaded side; do not bring
deposit to base of projection

Punch: Do not coat entire end; heaviest
deposit where wear will be highest

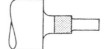

Journal: Taper deposit to minimize stress
concentration

 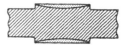

Shaft: Blend deposit; put on boss to prevent
shaft weakening

FIGURE II.8
Typical fusion hardfacing applications.

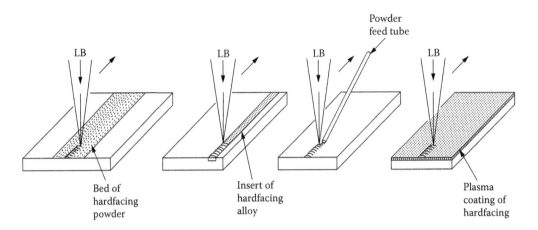

FIGURE II.9
Techniques for application of hardfacing deposits with laser welding.

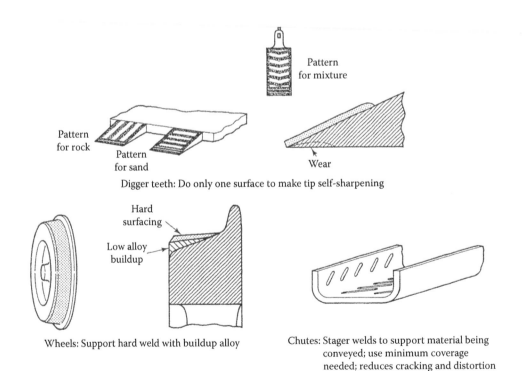

FIGURE II.10
Suggested fusion hardfacing bead profiles.

Appendix III: Thermal Spray Processes and Consumables

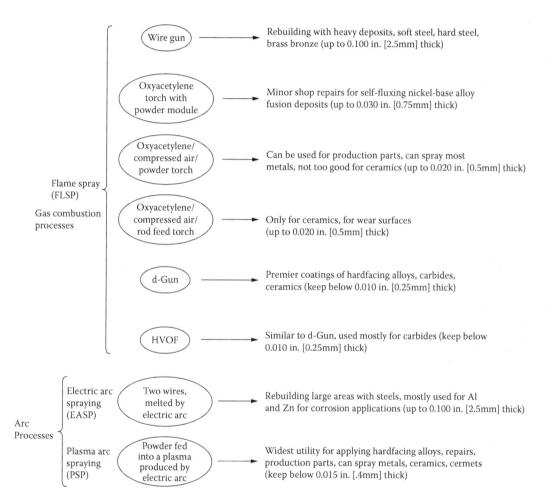

FIGURE III.1
Application guidelines for thermal spray processes.

Commonly Used Nonfusion Thermal Spray Consumables

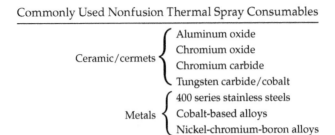

Ceramic/cermets
- Aluminum oxide
- Chromium oxide
- Chromium carbide
- Tungsten carbide/cobalt

Metals
- 400 series stainless steels
- Cobalt-based alloys
- Nickel-chromium-boron alloys

FIGURE III.2
Commonly used nonfusion hardfacing consumables.

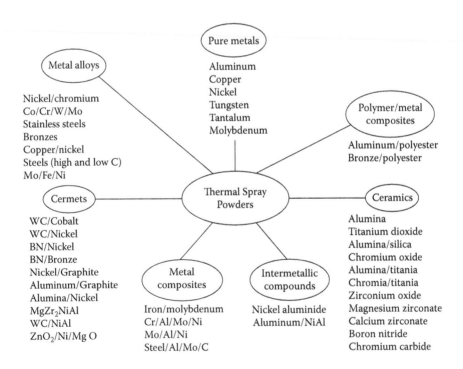

Metal alloys
Nickel/chromium
Co/Cr/W/Mo
Stainless steels
Bronzes
Copper/nickel
Steels (high and low C)
Mo/Fe/Ni

Pure metals
Aluminum
Copper
Nickel
Tungsten
Tantalum
Molybdenum

Polymer/metal composites
Aluminum/polyester
Bronze/polyester

Cermets
WC/Cobalt
WC/Nickel
BN/Nickel
BN/Bronze
Nickel/Graphite
Aluminum/Graphite
Alumina/Nickel
MgZr$_2$NiAl
WC/NiAl
ZnO$_2$/Ni/Mg O

Thermal Spray Powders

Metal composites
Iron/molybdenum
Cr/Al/Mo/Ni
Mo/Al/Ni
Steel/Al/Mo/C

Intermetallic compounds
Nickel aluminide
Aluminum/NiAl

Ceramics
Alumina
Titanium dioxide
Alumina/silica
Chromium oxide
Alumina/titania
Chromia/titania
Zirconium oxide
Magnesium zirconate
Calcium zirconate
Boron nitride
Chromium carbide

FIGURE III.3
Spectrum of powder consumables for thermal spray processes.

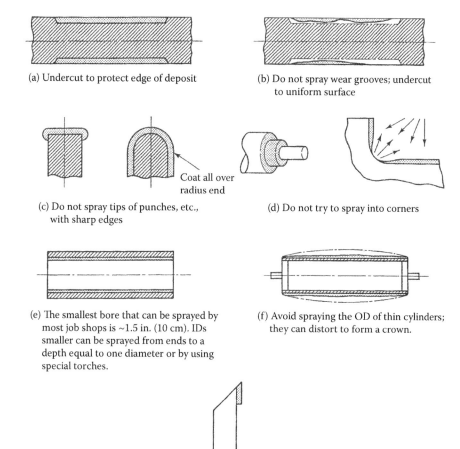

(a) Undercut to protect edge of deposit

(b) Do not spray wear grooves; undercut to uniform surface

Coat all over radius end

(c) Do not spray tips of punches, etc., with sharp edges

(d) Do not try to spray into corners

(e) The smallest bore that can be sprayed by most job shops is ~1.5 in. (10 cm). IDs smaller can be sprayed from ends to a depth equal to one diameter or by using special torches.

(f) Avoid spraying the OD of thin cylinders; they can distort to form a crown.

(g) Avoid grinding of sprayed coatings to form cutting edges; only use as-sprayed and very thin (0.0003 to 0.0005 in.; 7.5 to 12.5 mm).

FIGURE III.4
Thermal spray application suggestions.

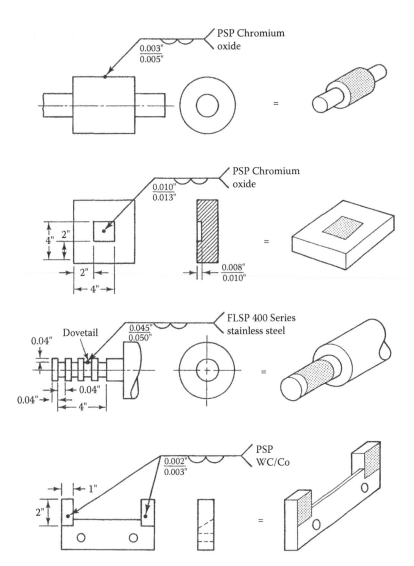

FIGURE III.5
Using the AWS system to specify a thermal spray coating.

Appendix IV: Diffusion Treatments

Comparison of Some Commonly Used Diffusion Treatments

Process	Suitable Substrates	Typical Case Thickness	Quench Required
Carburizing	Low-carbon and alloy steels	5 to 50 mils (0.12 to 1.25 mm)	Yes
Carbonitriding	Low-carbon and alloy steels	2 to 10 mils (50 to 250 μm)	Yes
Ferritic nitrocarburizing	Low-carbon and alloy steels	0.1 to 0.5 mil (2.5 to 12 .5 μm)	Yes
Nitriding	Nitriding steels and some alloy steels	2 to 20 mils (50 μm to 0.5 mm)	No

FIGURE IV.1
Process comparison.

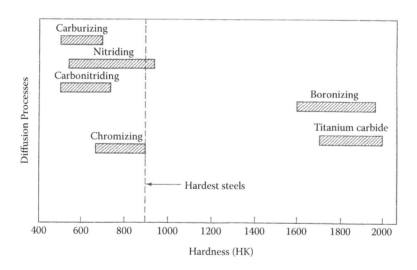

FIGURE IV.2
Typical hardnesses for various diffusion treatments.

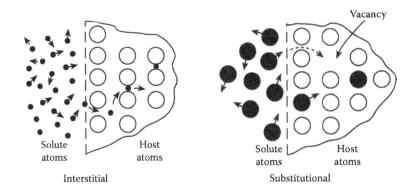

FIGURE IV.3
Schematic of a diffusion treatment.

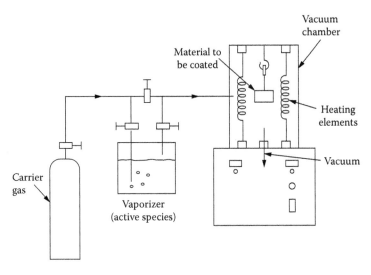

FIGURE IV.4
Schematic of CVD coating.

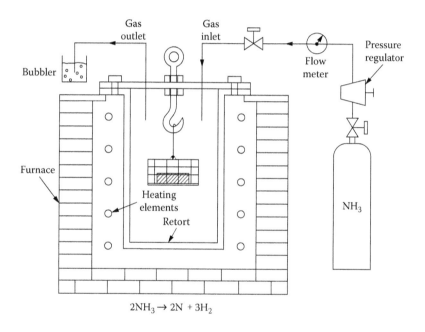

FIGURE IV.5
Schematic of gas nitriding.

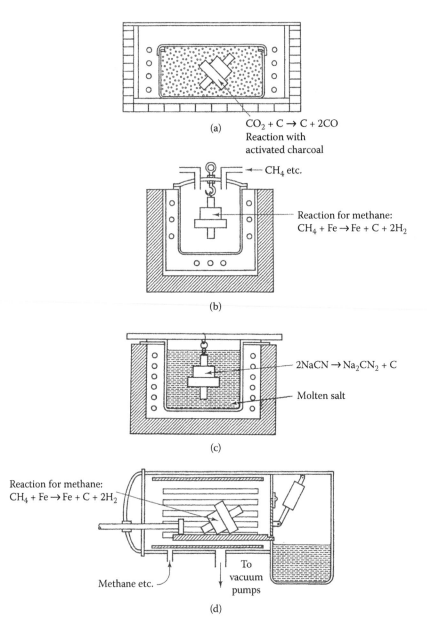

FIGURE IV.6
Various carburizing processes; a= pack, b = gas, c = salt, d = vacuum.

Appendix V: Selective Hardening

Materials That Are Commonly Flame and Induction Hardened and Their Normal Hardness Ranges

Carbon Steels (HRC)		Alloy Steels (HRC)		Tool Steels (HRC)		Cast Irons (HRC)	
1025–1030	(40–45)	3140	(50–60)	01	(58–60)	Meehanite GA	(55–62)
1035–1040	(45–50)	4140	(50–60)	02	(56–60)	Ductile, 80–60–03	(55–62)
1045	(52–55)	4340	(54–60)	S1	(50–55)	Gray	(45–55)
1050	(55–61)	6145	(54–62)	P20	(45–50)		
1145	(52–55)	52100	(58–62)				
1065	(60–63)						

FIGURE V.1
Materials that are commonly flame and induction hardened.

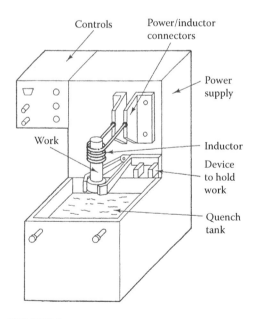

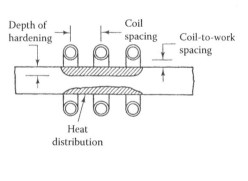

FIGURE V.2
Typical induction heating system.

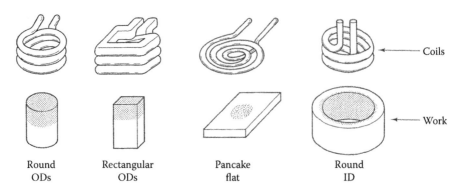

FIGURE V.3
Typical coil configurations for induction hardening.

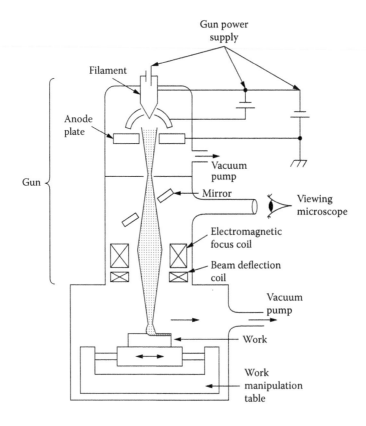

FIGURE V.4
Schematic of an electron beam system that can be used for selective hardening.

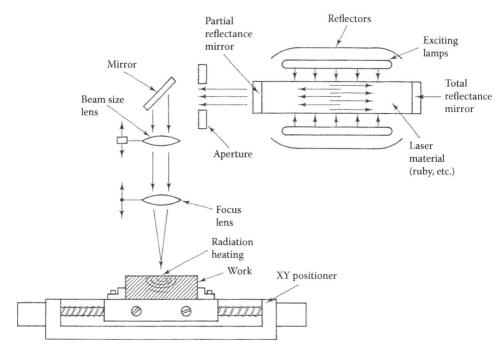

FIGURE V.5
Schematic of selective hardening with a laser.

Appendix VI: Thin Coatings and Surface Treatments

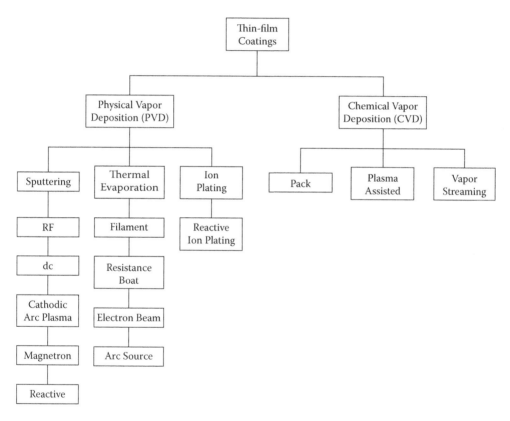

FIGURE VI.1

The spectrum of processes that are used for thin coatings.

Thermal Evaporation

Pure Metals	Precious Metals
Aluminum	Platinum
Iron	Palladium
Cobalt	Rhodium
Copper	Gold
Nickel	Silver
Cadmium	*Alloys*
Silicon	
Germanium	Brass
Tin	Inconel
	MCrAlYs
Refractory Metals	*Ceramics (Metal Compounds)*
Chromium	
Tungsten	Silicon oxide
Molybdenum	Tantalum oxide
Tantalum	Titanium oxide
	Aluminum oxide
	Magnesium fluoride

FIGURE VI.2
Materials that can be applied by thermal evaporation.

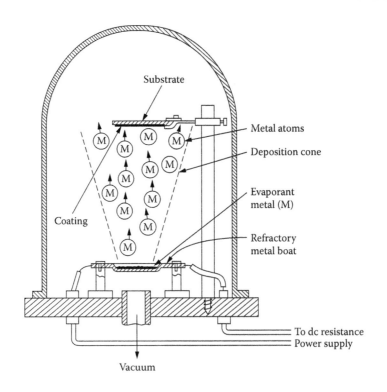

FIGURE VI.3
Schematic of thermal evaporation.

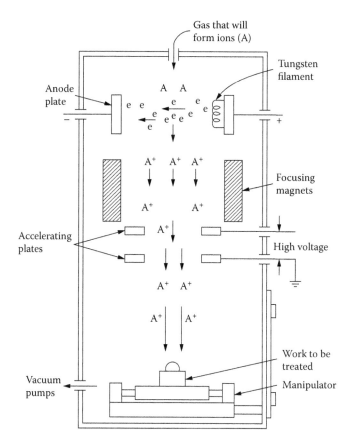

FIGURE VI.4
Schematic of ion implantation.

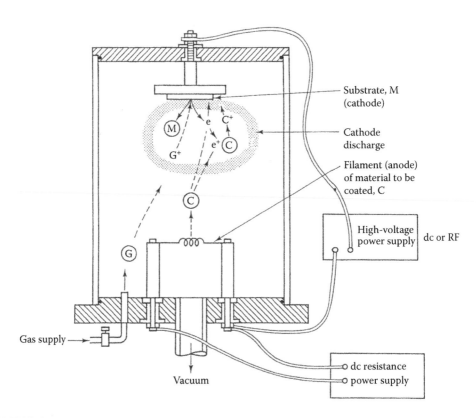

FIGURE VI.5
Schematic of ion plating.

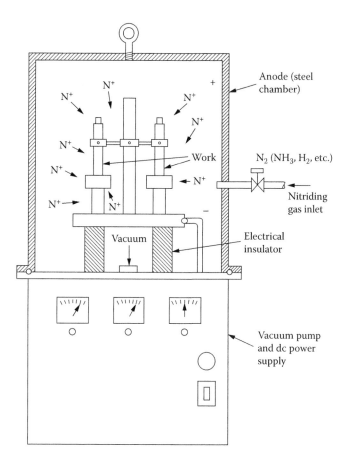

FIGURE VI.6
Schematic of ion nitriding.

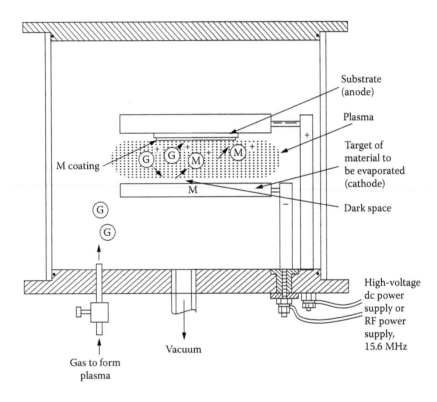

FIGURE VI.7
Schematic of planar diode sputter deposition.

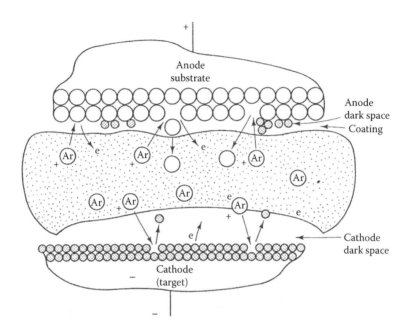

FIGURE VI.8
Typical reactions in sputter coating and coating cross section (250×).

Appendix VII: Plated and Conversion Coatings

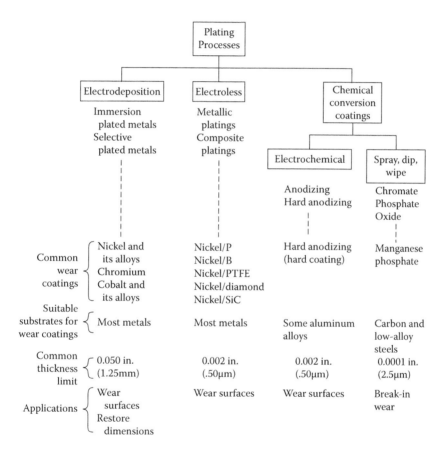

FIGURE VII.1
Plating processes that have utility in tribology.

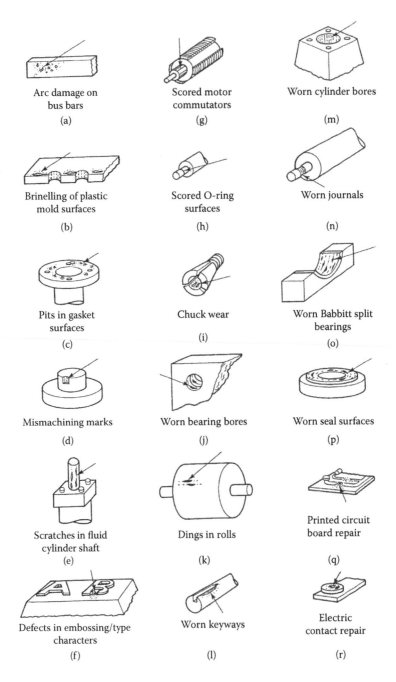

Arc damage on
bus bars
(a)

Scored motor
commutators
(g)

Worn cylinder bores
(m)

Brinelling of plastic
mold surfaces
(b)

Scored O-ring
surfaces
(h)

Worn journals
(n)

Pits in gasket
surfaces
(c)

Chuck wear
(i)

Worn Babbitt split
bearings
(o)

Mismachining marks
(d)

Worn bearing bores
(j)

Worn seal surfaces
(p)

Scratches in fluid
cylinder shaft
(e)

Dings in rolls
(k)

Printed circuit
board repair
(q)

Defects in embossing/type
characters
(f)

Worn keyways
(l)

Electric
contact repair
(r)

FIGURE VII.2
Typical applications of selective plating.

Appendix VIII: Properties of Engineering Materials

Chemical Symbols and Properties of Some Elements

Element	Symbol	Melting Point °F	(°C)	Density, g/cm³	Resistivity at 20°C, 10⁻⁶ ohm-cm	Linear Coefficient of Thermal Expansion,[f,g] 10⁻⁶ in./in.°F[e]	Thermal Conductivity at 25°C, W/cm/°C[h]
Silver	Ag	1761	(960)	10.49	1.59	10.9	4.29
Aluminum	Al	1220	(660)	2.70	2.66	13.1	2.37
Gold	Au	1945	(1062)	19.32	2.44	7.9	3.19
Beryllium	Be	2345	(1285)	1.84	4.20	6.4	2.01
Carbon	C	6740[e]	(3726)	2.25	75.00	0.3–2.4	2.1
Calcium	Ca	1564	(851)	1.54	4.60[a]	12.4	2.01
Columbium	Cb[d]	4474	(2467)	8.57	14.60	4.06	0.537
Cerium	Ce	1463	(795)	6.66	75.00[a]	4.44	0.113
Cobalt	Co	2719	(1492)	8.90	5.68[b]	7.66	1.00
Chromium	Cr	3407	(1825)	7.19	12.80	3.4	0.939
Copper	Cu	1981	(1082)	8.94	1.69	9.2	4.01
Iron	Fe	2795	(1535)	7.87	10.70	6.53	0.804
Germanium	Ge	1717	(936)	5.32	60×10^6	3.19	
Hafnium	Hf	4032	(2222)	13.29	35.5	3.1	0.230
Mercury	Hg	−70	(−38)	13.55	95.78		0.083
Iridium	Ir	4370	(2410)	22.42	5.30[b]	3.8	1.47
Lanthanum	La	1688	(892)	6.17	57.00[a]	3.77	0.134
Magnesium	Mg	1204	(651)	1.74	4.46	15.05	1.56
Manganese	Mn	2271	(1244)	7.44	185	12.22	0.078
Molybdenum	Mo	4730	(2610)	10.22	5.78[a]	2.7	1.38
Nickel	Ni	2646	(1452)	8.90	7.8	7.39	0.909
Osmium	Os	5432	(3000)	22.50	9.5[b]	2.6	0.076
Lead	Pb	621	(327)	11.34	22	16.3	0.353
Palladium	Pd	2826	(1552)	12.02	10.3	6.53	0.718
Platinum	Pt	3216	(1768)	21.40	10.58	4.9	0.716
Plutonium	Pu	1183	(639)	19.84	146.45[b]	30.55	0.067
Rhenium	Re	5755	(3179)	21.02	19.14	3.7	0.480
Rhodium	Rh	3560	(1960)	12.44	4.7[b]	4.6	1.50
Silicon	Si	3520	(1382)	2.33	15×10^{6c}	1.6–4.1	1.49
Tin	Sn	449	(232)	7.30	11.5	13	0.668
Tantalum	Ta	5425	(2996)	16.60	13.6[a]	3.6	0.575
Thorium	Th	3182	(1750)	11.66	18[a]	6.9	0.540
Titanium	Ti	3035	(1668)	4.54	42	4.67	0.219
Uranium	U	2070	(1132)	19.07	30	3.8–7.8	0.275
Vanadium	V	3486	(1918)	6.11	24.8	4.6	0.307

continued

Chemical Symbols and Properties of Some Elements (continued)

Element	Symbol	Melting Point °F	(°C)	Density, g/cm³	Resistivity at 20°C, 10^{-6} ohm-cm	Linear Coefficient of Thermal Expansion,[f,g] 10^{-6} in./in.°F[e]	Thermal Conductivity at 25°C, W/cm/°C[h]
Tungsten	W	6170	(3410)	19.30	5.5	2.55	1.73
Yttrium	Y	2748	(1508)	4.47	65		0.172
Zinc	Zn	787	(419)	7.13	5.75	22	1.16
Zirconium	Zr	3366	(1852)	6.45	44	5.8	0.227

[a] At 25°C.
[b] At 0°C.
[c] At 300°C.
[d] Also called niobium, Nb.
[e] Sublimes.
[f] Multiply by 1.8 to convert to cm/cm°C.
[g] At 68°F (20°C).
[h] Multiply by 55.7 to convert to Btu/ft/°F.

FIGURE VIII.1 (continued)
Chemical symbols for elements and some elemental properties.

Compatibility of Various Surface-Hardening Processes with Important Metal Systems

Substrate	Applicable Fusion Hardfacings	Applicable Other Processes
Aluminum alloys	None	Anodizing, electroplate, PVD sputter coatings, thermal spray
Copper alloys	CuZn CuAl NiCrB	Plating, PVD, CVD coatings, ion plating, thermal spray
Low-carbon steels	All	Diffusion processes, thin-film coatings, platings, thermal spray
Alloy steels (hardenable)	All, but weldability concerns	Selective hardening, plating, thin-film coatings, high-energy modifications, nitriding
Cast irons	Poor arc weldability OAW: Bare rods	Selective hardening, plating, thermal spray
Martensitic stainless steels	Poor arc weldability OAW: Bare rods	Selective hardening, plating, thin films, nitriding, thermal spray
Austenitic stainless steels	All	Nitriding, thin films, plating, thermal spray
Magnesium alloys	None	Anodizing, PVD sputter coating, thermal spray
Nickel alloys	All	Thin films, plating, thermal spray
Titanium alloys	None	Anodizing, thin films, thermal spray
Zinc alloys	None	Plating, thermal spray
Tool steels	None	Thin films, plating, thermal spray

FIGURE VIII.2
Surface-hardening processes with metal systems.

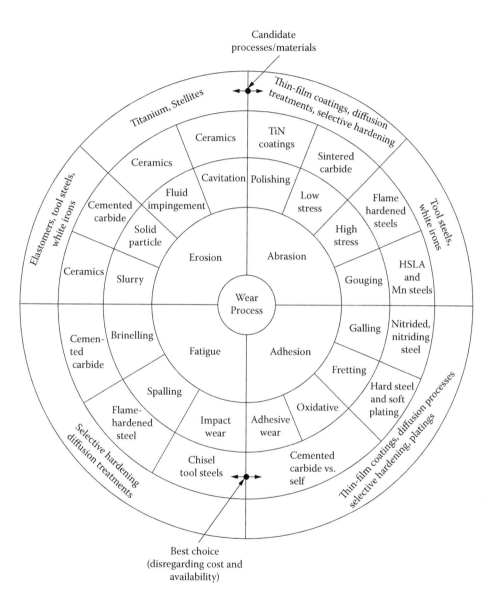

FIGURE VIII.3
Some candidate materials for various types of wear and erosion.

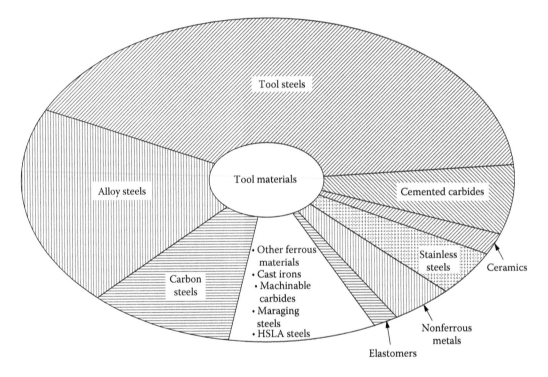

FIGURE VIII.4
Estimated relative usage of engineering materials for friction, wear, and erosion applications.

Index

For Product Safety Concerns and Information please contact our EU representative GPSR@taylorandfrancis.com Taylor & Francis Verlag GmbH, Kaufingerstraße 24, 80331 München, Germany

Printed and bound by CPI Group (UK) Ltd, Croydon, CR0 4YY

01/05/2025

01858486-0001